I0469309

BASIC TOOLS AND RESOURCES
FOR
FIRE INVESTIGATORS:
A HANDBOOK

FEDERAL EMERGENCY MANAGEMENT AGENCY
UNITED STATES FIRE ADMINISTRATION

September 1992

TABLE OF CONTENTS

Chapter 1
OVERVIEW

PURPOSE OF THIS HANDBOOK

This handbook is designed to serve as a compendium of basic information about the tools and techniques of fire investigation. The goal here is to provide a breadth of useful information in as concise and as accessible a package as possible. To do this, the scope has been limited to the tools and techniques associated with determining and verifying the origin and cause of accidental and incendiary fires. This is not a law enforcement procedures manual, information about surveillance, criminal identification, interview and interrogation procedures, insurance fraud, and procedures related to investigating other criminal activities often associated with arson are not described here. Information about presentation of fire-related evidence in criminal and civil trials is only discussed in the context of how it relates to the advantages and disadvantages of specific techniques.

RESEARCH METHOD

The information in this handbook has been assembled from a wide array of sources, including the fire investigation and forensic science literature, interviews with experts in fire investigation and forensic science practitioners, and the nationally recognized standards of practice for fire investigation and forensic science. Members of both professions from across the country, and in one case another country - Australia - have lent their considerable experience and expertise to this project by responding to published requests for information and participating in interviews about what they do and how they do it.

Published Request Responses

The published requests for information were circulated to all of the major fire service, fire investigation, code enforcement, and fire protection publications asking investigators to respond to the following questions:

- What basic tools are issued to all your investigators? (Please provide lists)

- What equipment not currently available to you do you most need?? (Please identify)

- Do you currently use canines in fire investigation? Are they certified? Have their findings held up in court?

- What type of gas chromatograph do you use? How has it performed?

- Where do you send your evidence - to your own lab, the police department lab, a state lab or a private lab?

Useful responses were received from 29 fire investigation units. These agencies, shown on page 2, represented all types of fire investigation units, career and volunteer, from a diverse range of communities large and small, urban, suburban, and rural. Agencies from different levels of government responded as well, including state fire marshal's offices, county fire investigation units, and local fire investigators. Two encouraging results were apparent from the nature and diversity of the responses. First, that despite their differences, all of these jurisdictions had many things in common about the way they conducted fire investigations and what they used to do this. The other was just as important - that despite the similarities, there was no single way to do everything.

These responses provided valuable insight into the administrative environment in the fire investigation field.

Interviews

To corroborate and further expand on the findings of the publication responses, a series of interviews with local, state, and federal fire investigators and laboratory personnel from across the nation was conducted. The agencies involved are shown on page 2. These interviews provided additional insight and valuable background information about the use of technology, disputes about techniques, and the advantages and disadvantages of various methods or procedures.

As a result of these responses, this handbook tries to do three things:

- Explain what is out there and how it is used;

- Explain how much things cost and what advantages, disadvantages, costs, and benefits are associated with the tools and techniques described;

- And, explain what constitutes a good basic or minimum complement of tools, equipment, and technical capability.

1

Fire Investigation Units
Responding to Request for Information

Allentown Fire Department, PA
Atlantic County Prosecutor's Office, NJ
California State Fire Marshal's Office, CA
Chino Valley Fire Department; CA
Cincinnati Division of Fire, OH
Clark County Fire Department, NV
Colerain Township Fire Department, OH'
District Six Volunteer Fire Dept., Baton Rouge, LA
Gainesville Fire/Rescue Department, FL
Hall County Public Safety Department, GA
Hartford Fire Department, CT
Kent Fire Department, WA
Lewis County Fire Investigation Unit, NY
Maryland State Fire Marshal's Office, MD
Metropolitan Fire Brigades Board, Melbourne, Australia
Mokena Fire Protection District, IL
Moss Township Fire Department, MS
Orange County Fire Department, CA
Palm Bay Fire Department, CA
Piqua Fire Department, OH
Riley Township-Pandora Volunteer Fire Dept., OH
Rogers Fire Department, AR
San Mateo Arson Task Force, CA
Sellersburg Fire Department, IN
South Kitsap Fire Department, WA
Ventura County Fire Protection District, CA
Vineland Fire Department, CT
Windsor Fire Department, CT
Wyoming Dept. of Fire Prevention and Electrical Safety, WY

Agencies Interviewed

• Arkansas Department of State Police, Fire Marshal Section
• U. S. Department of the Treasury, Bureau of Alcohol, Tobacco, and Firearms
• Connecticut State Police, Canine Section
• Eastern Kentucky University, Department of Loss Prevention and Safety
• Florida Division of State Fire Marshal
• Insurance Committee for Arson Control
• International Association of Arson Investigation, Forensic Science and Training and Education Committees
• Minnesota State Fire Marshal Division
• Nebraska State Fire Marshal's Office
• Oklahoma City Fire Department, OK
• Pennsylvania State Police, Fire Marshal Division
• Portland Fire Bureau, OR
• Wyoming Dept. of Fire Prevention and Electrical Safety

SUMMARY OF FINDINGS:

Basic Tools and Equipment

Fire investigators are using a wide variety 'of tools and equipment in their work. Many of these are commonplace, others quite unique and sophisticated. Many departments have either acquired or desire electronic equipment like video cameras, cellular telephones, facsimile machines, and personal computers. A decided trend throughout the public-safety field which was not strongly demonstrated by the publication responses but reinforced during interviews, was the need for personal protective equipment for fire investigators. Several fire investigation professionals pointed out the riced for respiratory protective equipment, and one noted the need for latex gloves for preventing contact with blood-borne pathogens. Yet another office discussed the need for Hepatitis B vaccinations for its employees. Many agencies employed sworn law enforcement officers who were issued firearms, ammunition, handcuffs, and ballistic vests.

Accelerant Detection

A traditional law enforcement approach - the use of canines for detection purposes - is gaining popularity for accelerant residue detection. Hand-held electronic "sniffers" remain popular as well. Some agencies use both techniques in tandem to increase their sampling confidence. Both techniques have their proponents. Both have become well-established means of identifying accelerant residue samples for collection and analysis. Likewise, both have been accepted in court. So far, no one is reporting any problems with canine evidence in court proceedings. However, many forensic science professionals caution that if rigorous training and certification standards are not established and maintained, this excellent record may be in jeopardy.

Collection; Packaging, and Analysis of Evidence

Much of the early attention here focused on laboratory techniques, but as work progressed it became evident that procedures for collecting, packaging, and storing evidence were equally if not more important. Analysts interviewed during the development of this manual and published articles in the literature emphasized that evidence which is not collected or which is collected or handled improperly will severely compromise any investigation. Since no clear guidelines exist on what evidence or how many samples to collect, interest shifted to determining if guidelines were discernable from the current practices in the field. Advances in packaging technologies have introduced a few innovations, such as Kapak^{TM} bags which are resistant to vapor leakage, but the best evidence containers for accelerant residue samples remain clean metal paint cans with friction lids. Beyond collection and packaging, the nature of the fire scenario and, the types of evidence submitted to the laboratory determine how it is processed by the forensic science technicians. The effectiveness of, an analysis is heavily influenced by communication between the field investigator and the laboratory analyst.

Computers

The use of computers in the fire service has been growing as elsewhere in business and government. Federal efforts to develop useful computer-aided tools for fire investigators began in the early 1980s with the United States Fire Administration's development of the Arson Information Management System (AIMS). Literature research and interviews revealed that a large number of commercial products exist which, follow the AIMS lead. Also, a wide variety of application programs exist for word processing, database management, graphics, communications, and numerical analysis. As these programs have become more "user-friendly" they have become more popular among fire investigators. Investigators can use most of the popular commercial programs for writing reports, preparing fire scene diagrams, maintaining investigation records, analyzing financial information and fire statistics, and communicating with other fire protection and law enforcement professionals via electronic information services. The horizon for use of computers in the fire service continues expanding with the introduction of increasingly sophisticated computer fire models for analyzing complex fire protection problems. Programs like the National institute for Standards and Technology, Building and Fire Research Laboratory's (NIST/BFRL) HAZARD I model are designed to be easier learn and use. These programs offer fire investigators a set of tools for understanding and documenting what happened in fires they investigate.

Information Resources

All complex, fields of knowledge are under continuous revision and interpretation. Fire investigation and fire science are certainly no exceptions. Keeping well-informed and current is a constant challenge. Fortunately, fire investigators have access to a wide array of information resources, including periodicals, books and texts, codes and standards, on-line information services, and conference proceedings. The federal government maintains: a large number of reference sources for fire investigators, including the Fire and *Arson Investigators' Field Index Directory, Arson Resource Directory, Establishing an Arson Strike Force,* and *A View of Management in Fire Investigation Units.*

ORGANIZATION OF THIS REPORT

The authors chose to organize this handbook much the same way a good investigator conducts an origin and cause inquiry. The fire investigator works from the least burned or damaged area to the most heavily damaged area in a systematic fashion, collects information and evidence, processes the information, analyzes the results and presents his or her case to a prosecutor and/or court. This handbook works from the hardware, tools, and equipment used at the fire scene to the scientific and technical information and analysis necessary to resolve a complex case.

Chapter 2
FIELD INVESTIGATION TOOLS

The most important fire investigation field tools are keen observational skills, innate curiosity, and an interest in determining the truth. This section details the hardware required to put the basic tools of intellect, observation, and perseverance to work in a difficult and challenging environment - the fire scene. The utility of all of these tools and equipment can be expanded by proper training and use.

This discussion includes accidental and incendiary alike. Some of the tools discussed are considered useful in determining the origin or cause of either incendiary or accidental fires, most are useful in investigating both types of fires.

HAND TOOLS AND BASIC EQUIPMENT

Fire investigators must perform many of the tasks associated with overhaul in the process of identifying and eliminating potential ignition sources, collecting evidence, and determining the mechanisms of fire behavior and fire spread. Some simple hand tools and other equipment should be carried by or be available to fire investigators to perform tasks such as disassembling equipment, opening packages, removing interior finish and trim, and obtaining evidence samples. The following sections detail the equipment investigators say they use to perform many routine fire investigation tasks. These items include tools for cutting, drilling, prying, loosening, and measuring. A complete list of these tools appears in Appendix A.

EVIDENCE COLLECTION EQUIPMENT

Evidence collection and preservation is one of the most important tasks associated with fire investigation. Everything needed to confirm the cause, origin, or contributing factors associated with a particular fire must be collected and catalogued. In the case of incendiary and suspicious fires, evidence about suspects and people associated with the premises or the fire scene must be collected and preserved as well.

Essential Equipment

Key tools associated with these tasks include equipment for photographing, collecting, marking, and preserving fire scene debris samples, materials for documenting and diagramming the overall condition of the scene, and tools for collecting the statements and accounts of witnesses. A list of basic evidence collection equipment appears in Appendix A. Many of the items on that list are quite costly. Some departments with limited budgets will find it necessary to limit their equipment to the bare essentials. Some of the basic evidence collection equipment may include a simple 35-mm automatic camera, a small assortment of evidence containers, a measuring tape or rule, and drawing and writing supplies.

Desirable Equipment

Many agencies have acquired video camcorders, videocassette recorders, and videotape editing machines or placed them high on their "wish lists." Video and advanced telecommunications are two of the fastest growing capabilities of most fire investigation units. Although sophisticated high-resolution video camcorders capable of operating in low-light-level conditions costs approximately $2,000, a reasonable videotape camcorder and playback unit can be obtained for less than $1,200. Videotape capability has become so popular so quickly because it provides the investigator with a tool for capturing action and documenting the fire scene while maintaining spatial and temporal perspective. Moreover, when the fire investigator can arrive at the scene before the fire is extinguished, videotape is afar superior method of capturing live action fire sequences for reconstructing the fire scenario later.

The use of videotape as a law enforcement and investigation tool has become so commonplace that many law enforcement agencies have begun mounting videotape cameras in their vehicles to record routine traffic stops. Following suit, some fire departments are considering installing video camcorders in fire apparatus to record fire operations. Likewise, many fire investigation units use video camcorders to routinely record the crowds assembled at fire scenes to identify possible arson suspects. All of these applications are significantly enhanced by editing capability. Raw videotape can be edited and captioned to reinforce key points for training purposes or to highlight key points when used for presentation at trial.

PERSONAL AND PROTECTIVE EQUIPMENT

Most fire investigators assigned to fire service or law enforcement units will need some basic equipment for their personal use and protection. Items like badges, ID cards, business cards, and uniforms help identify them to the public, which can cut two ways: it can help protect them or identify them as a target. Items like flashlights and clipboards are basic tools for most investigators as well. Fire scenes are inherently dangerous places to work; therefore, protective equipment should be provided to minimize the risk of personal injury. A list of personal and protective equipment appears in Appendix A.

Personal Protective Equipment

In recent years, increased awareness of the hazards to firefighters have produced significant improvements in personal protective equipment. However, the hazards to fire investigators are often given inadequate consideration. Just because the fire is out does not mean the hazard has passed. Toxic combustion products like carbon monoxide, carbon dioxide, cyanides, sulphur dioxide, hydrogen sulfide, acrolein, nitrogen oxides, and other substances exposed by the fire such as asbestos are just a few of the toxic inhalation hazards present on the fireground after a fire is extinguished. Fire weakened and water-logged structures present the continuing danger of collapse; and, sharp objects and deformed materials may present trip, fall, or abrasion hazards. In short, the fireground only becomes marginally less hazardous after the fire is out, and fire investigators must recognize and protect themselves from many of the same hazards as firefighters. These include the hazards of acute exposures which may incapacitate an investigator immediately and those which produce chronic or long-term health effects which produce gradual debilitation.

Many investigation units now issue their investigators half-mask respirators and air monitoring equipment to guard against fireground inhalation hazards. The most popular filters for use with these respirators are designed to filter out volatile organic chemicals, acid vapors, and particulate. Air monitoring equipment is used to detect and alarm upon the presence of diminished oxygen concentration or the presence or elevated carbon monoxide levels.

Coveralls, steel toe, steel shank shoes or boots, hardhats or helmets, firefighter turnout clothing, gloves, and goggles complete the fire investigator's protective ensemble. These items prevent injuries from abrasion, impact, and puncture. A list of protective equipment appears in Appendix A.

Law Enforcement Equipment

Some fire investigators have dual authority to investigate fires and make criminal arrests. Different duties come with the added authority of sworn law enforcement officers. These factors also influence how the fire investigators, the public, and potential suspects may interact. Most law enforcement officers not only possess the right to bear firearms, but a duty to carry these weapons on and off-duty as well. The risks associated with carrying and using firearms bear special consideration for fire investigators. The issuance and use of ballistic vests should be seriously considered anytime fire investigators will USC or carry firearms, as well as when they expect them to be used against them. To further minimize the risks of bodily injury to them or members of the public, most law enforcement agencies require suspects to be restrained when taken into custody. As such, handcuffs or other restraint devices should be issued when the fire investigator's authority and duties include arrest powers.

COMMUNICATIONS EQUIPMENT

The acquisition of communications equipment has become one of the top priorities of many fire investigation agencies. The reasons is clear: access to the right communications equipment saves a fire investigator time and enhances his or her ability to do a good job. With the costs dropping rapidly, most of these tools are now within easy reach of most public agencies. However, while the costs of acquiring tools like cellular telephones or facsimile machines may be relatively low, the cost of using and maintaining such services is usually higher than standard telephone service. Agencies should become familiar with the costs of operating cellular telephones, pagers, and facsimile machines to prevent unwelcome surprises.

FIRE INVESTIGATION VEHICLES

Many agencies have found it desirable to have a dedicated vehicle for transporting the wide array of tools and equipment they require to fire Scenes. Often these vehicles also serve as command posts. Like other fire service or law enforcement vehicles, these "arson vans" are often highly specialized. Usually, a fire investigation field unit consists of a van or truck with an enclosed cargo body. Most are equipped with emergency response equipment, public address systems; and radios. Field units of this type permit fire investigators to bring a wider variety of tools and equipment to the scene than would be possible with a staff car. Some of these items include digging tools, salvage tools, and scene security supplies. Some units are equipped with cellular telephones, cellular facsimile machines, portable computers, remote surveillance equipment, and forensic analysis equipment. The configuration of a local fire investigation field unit should be tailored to the jurisdiction's particular needs and budget. State motor vehicle laws dictate the types of warning equipment required if the unit will be used to respond to emergencies.

Chapter 3
ACCELERANT DETECTION

One of the principal techniques for evaluating whether a fire is incendiary is to test for the presence of flammable or combustible liquid accelerants. This section presents the most widely used techniques for detecting the presence *(not the concentration)* of flammable or combustible liquid accelerants in fire debris. Regardless of the investigative technique employed - analytic equipment, accelerant detection canines, or the investigator's senses of sight, touch, and smell - its reliability is dependent upon experience and judgement.

OVERVIEW

Eight types of field tests or instruments are recognized as valid for detecting the presence of flammable liquid vapors at fire scenes. However, only five of these methods have proven practical enough to become widely used on the fireground.

Many fire investigation agencies use mechanical or electrical devices for detecting the presence of flammable liquid accelerant vapors. Although four types of these devices, sometimes referred to as "sniffers," have proven reliable and widely popular, each has distinctive reliability drawbacks. "Portable" versions of gas chromatographs and spectrophotometers have been tried in the field, but have proven to unstable to be reliable.

Chemical indicators which change color in the presence of specific hydrocarbon compounds were tried in the field but found impractical. Unfortunately, this method is often too inaccurate or unspecific for fire investigation needs and is no longer widely used.

Human olfactory testing, is the oldest and most widely practiced accelerant detection technique. It involves the investigator smelling fire debris for the characteristic odor of a flammable solvent. The major drawback of this method is that the human sense of smell becomes fatigued after prolonged exposure to certain chemicals. However, in recent years the introduction of accelerant detection canines has vastly improved the reliability and performance of olfactory analysis at fire scenes. Trained dog and handler teams have demonstrated dramatic improvements in consistency, reliability, and performance in accelerant detection as verified by forensic analysis. This section examines the use of mechanical or electrical detection instruments and the use of accelerant detection canines. Although one form of detection may appear preferable to another, no detection method is 100% reliable or accurate. Many agencies,

therefore, prefer to use more than one field detection technique to improve their sampling "hit" rates.

MECHANICAL AND ELECTRONIC EQUIPMENT

Mechanical or electrical devices for detecting the presence of accelerant vapors fall into four major categories: catalytic combustion detectors, photo-ionization detectors, Toguchi sensor or semiconductor-based instruments, and flame ionization detectors. Detectors based on the catalytic combustion principal have traditionally been the most common and widely used instruments for detecting the presence of accelerants, but both photo-ionization and semiconductor-based instruments have become increasingly popular because of their improved sensitivity and selectivity. Flame ionization detectors are the least common of these field instruments.

Catalytic Combustion Detectors

A catalytic combustion detector consists of three basic components: a pump or other means of drawing a sample of vapor/air mixture into the instrument; a chamber containing a platinum wire coil which is part of a circuit known as a "Wheatstone Bridge"; and, a gauge for indicating the change in the resistance of the coil. These detectors operate on the simple electro-mechanical principal that heat and electrical resistance are related. When a sample of accelerant contaminated vapor/air mixture is drawn into the chamber, the heat from the coil quickly oxidizes it, causing a temperature rise. This increase in temperature increases the resistance of the coil, which is registered on the instrument's gauge. The specific heat of combustion for a given gas or vapor is peculiar to its molecular weight and concentration.

Most catalytic combustion detectors are calibrated to minimize interference from trace hydrocarbons normally present in background air samples. Gauge markings usually correspond to concentrations of a known flammable gas or vapor concentration, very often methane. Some of the more elaborate devices produce audible warnings when the detected concentration exceeds certain proportions of the 'calibration gas's lower explosive limit. Since these devices are usually calibrated against a single known flammable gas or 'vapor as a standard, their reliability at predicting the concentration, of other materials or mixtures of material vapors of varying molecular weights is questionable. Nonetheless, most manufacturers supply conversion charts for translating instrument readings for the calibration standard into equivalent concentrations of other compounds.

Figure 3.1

Accelerant Detection Techniques

Test Method	Detection Sensitivity Range	Comment
Chemical color test	1 - 5 parts per 1,000	Least specific and sensitive detection or test method.
Catalytic combustion	1 ppm	Noise from pyrolysis products may interfere with results.
Photo-ionization detector	0.1 - 1 ppm	Good sensitivity and selectivity. Moisture and condensation do not affect sensitivity or selectivity, but may produce inaccurate output.
Semiconductor or Toguchi sensor-based instruments	1 - 5 ppm	Dual sensor devices have good sensitivity and selectivity, but may take more time to setup. Hydrocarbon residue accumulation may diminish sensitivity.
Flame ionization detector	1 - 5 ppm	May yield false "positive" results.
"Portable" gas chromatograph	1 - 5 ppm	Very expensive, not totally portable, requires calibration after setup, lengthy analytic procedure.
"Portable" infared spectrophotometry	1 - 5 ppm	Needs pure sample. Impractical for field use.
Olfactory test	1 ppb	Human olfactory sense can suffer from fatigue. Canine sensitivity and reliability dramatically superior.

Source: Lee, H. C. and Messina, D. A., *Evaluation of Arson Canine Testing Program,* Meriden, CT: Connecticut State Police, Forensic Science Laboratory.

The simplicity of catalytic combustion detectors, ease of operation, and widespread use in other emergency services applications such as detecting natural gas or liquified petroleum gas leaks have made them the most popular and widely used type of accelerant detection instrument. Many of these instruments also now feature solid state electronics, digital readouts, print recorders, and oxygen sensors.

Photo-ionization Detectors

Photo-ionization detectors use a high frequency light source in the far ultraviolet range to produce ionize contaminants in air samples entering the detector. When hydrocarbon samples are ionized, the free electrons are attracted to a cathode. The resulting change in current is amplified and registered by a device like a meter or strip chart recorder.

These devices are usually quite sensitive and exhibit good selectivity for hydrocarbon vapors. Photo-ionization detectors are considered to be up to ten times more sensitive than flame-ionization detectors, and are much more stable. The stability of the photo-ionization detector is a product of the energy source and power supply. Unlike the flame-ionization detector's hydrogen fueled flame, the battery powered ultraviolet light source produces a consistent and reliable energy source.

Moisture and temperature extremes may affect instrument readings. The presence of hydrocarbon pyrolysis products in the fire debris matrix may yield false positive readings, but perhaps no more so than other devices. An experienced operator can readily interpret the readings provided by the instrument by considering the sources analyzed.

These devices are small, very portable, and easy to purge, permitting rapid sampling of suspect areas. A good photo-ionization detector instrument costs approximately $4500.

Semiconductor or Toguchi Sensor Instruments

Instruments based on the Toguchi sensor employ a semiconductor chip to detect the presence of hydrocarbons vapors. Elevated hydrocarbon concentrations in air infiltrate the chip's crystalline structure, changing its conductive properties. These are registered and amplified to produce an analog reading on a meter. This simple design coupled with a duplicate sensor permits background sources to be eliminated through calibration or comparison.

One of the leading manufacturers of semiconductor based instruments for accelerant detection produces a device which employs dual sensor technology and multiple probes to permit the operator to zero-out background sources. This facilitates easy identification of pattern boundaries where liquid accelerants may have been poured to create trailers or fuses. Some devices also offer interchangeable sensors to permit variable selectivity for different applications.

Since hydrocarbons are permitted to enter the semiconductor's structure, they must burn-off before the device can be cleared to screen a new sample. Although this takes very little time in most cases, it can slow down even an experienced investigator. Moreover, some users complain that these devices become contaminated over time, making it increasingly difficult to eliminate background interference. Another serious limitation is that exposure to silicon vapors from certain sources such as brake fluid and some lubricating oils will destroy the sensor element.

A semiconductor-based instrument costs approximately $1200.

Flame Ionization Detectors

Flame ionization detectors mix the sample vapor/air mixture with gaseous hydrogen and burn the mixture. The high temperature flame produced by this combustion causes the molecules in the mixture to ionize; that is, break down into electrically charged chemical species. This causes the gas mixture to become electrically conductive. The degree of ionization, hence the degree of conductivity of the gas mixture, is measured using an electrometer to produce a reading indicating the presence of accelerants.

This technique is capable of detecting much smaller quantities of accelerant than catalytic combustion detectors and reporting them much more accurately. However, the sensitivity of the device often works against its reliability by producing false positive readings in the presence of pyrolysis products of certain materials like plastics and chemically treated wood. Another significant drawback is the need to carry a portable hydrogen fuel supply.

ACCELERANT DETECTION CANINES

The first accelerant detection canine program was developed in 1986 through a joint program of the Bureau of Alcohol, Tobacco, and Firearms, Connecticut State Police, Connecticut Bureau of State Fire Marshal, and New Haven (CT) State's Attorney's Office. Since that time, dozens of agencies nationwide have followed suit. Still, even ardent advocates of accelerant detection canine programs stress that the use of the dogs is just another weapon in the fire investigator's arsenal. They emphasize that training and support are the keys to successful use of this technique.

Training Programs

Many agencies and organizations have responded to meet the growing demand for accelerant detection canines. However, like any service, some programs and trainers are better and more credible than others. Aside from the technical differences in the training methods employed by each, the truest test of the credibility and performance of a canine training program is the results its dog/handler teams produce. The field performance of any program's graduates should be a significant consideration when evaluating and selecting a trainer for a new accelerant detection program. A key measure of their success is the percent of samples taken that prove accurate. The best programs claim a 97% "hit" rate for samples identified and submitted by accelerant canine teams for forensic analysis. Such high accuracy rates are obtained by rigorous training and recertification and a continual process of verification by forensic analysis.

The Connecticut State Police Canine Unit and the Maine State Police are two of the principal governmental agencies involved in training accelerant detection canines for law enforcement use. Both agencies offer training programs throughout the year for very reasonable fees. Each has demonstrated consistent success and proven results with its graduate dog/handler teams. These agencies may be contacted at the following addresses and phone numbers:

Canine Section
Bureau of State Fire Marshal
294 Colony Street
Mcriden, Connecticut 06540
(203) 236-6046

Maine Criminal Justice Academy
93 Silver Street
Waterville, Maine 04901
(207) 873-2651 or (207) 873-4691

Support

Effective use of accelerant detection canines is a team effort. Their success is founded in cooperation between law enforcement, fire investigation, forensic science, and criminal justice agencies. Agencies using accelerant detection canines stress that the dogs are just another tool for combatting arson. As a tool, their effectiveness is limited by the training and abilities of the trainer, handler, origin and cause investigator, law enforcement agency, forensic laboratory, and prosecutor responsible for investigating and prosecuting the case. The following general guidelines should help identify the types of support required to establish and maintain an effective accelerant detection canine program:

- Dogs and handlers should be trained as a team by competent professional canine trainers with a proven track record training law enforcement working dogs, preferably for specific use as accelerant detection canines.

- Accurate and contemporaneous records should be maintained of all dog and handler training, fire scene experience, and laboratory results of samples submitted for fire scenes and recertification or validation purposes.

- A rigorous daily training and work program should be established and followed to maintain and document the proficiency of the canine and handler.

- Accelerant detection canines should be used primarily to identify locations for collecting flammable or combustible liquid accelerant samples. Every location where an animal alerts to the presence of an accelerant should be sampled, and all samples should be submitted to a forensic laboratory for analysis and confirmation.

- Canines should be worked at every possible opportunity, accidental fire scenes included, to build experience and confidence. When canines work "negative" scenes, they should be reinforced afterwards with a "positive" training hit. This ensures that the dog is ready and willing to work every fire scene opportunity.

- Accelerant detection canine team evidence and testimony are credible and persuasive, but should never be relied on too heavily to build a case for prosecution. Canine teams should avoid letting their confidence in the dog's abilities persuade them to prosecute otherwise weak cases.

- Just because a dog does not alert at a scene does not mean the fire was not incendiary. Dogs are not perfect and neither are fire investigators. Every time a dog is used at a fire scene the investigator should have a clear understanding of what he or she expects to find or confirm. Negative and positive results alike must be evaluated carefully to determine what they mean to the fire investigation.

Successful use of this technique requires that each team member - canine handler, fire investigator, law enforcement professional, forensic scientist, and prosecutor - be knowledgeable about how the dog's use impacts their area of expertise and how it relates to their testimony at trial.

Chapter 4
COLLECTION, PACKAGING, AND ANALYSIS OF EVIDENCE

The successful presentation of fire scene evidence of incendiarism is often dependent upon the ability to document and verify the presence of accelerant residue in debris samples. Likewise, laboratory analysis often yields significant clues about the performance of fire protection features or factors contributing to fire growth and spread. A successful fire investigation program uses the scientific method at the fire scene and behind the scene - in the laboratory. This section discusses the procedures and equipment used in the forensic science community to conduct laboratory analysis of fire scene debris samples and photographs. The techniques discussed here will have applications to accidental as well as incendiary fires.

Only the fire investigator can determine what evidence needs to be collected to document a particular fire,. Techniques used to collect, package, and store evidence for forensic analysis are dictated by the types of evidence and analytic techniques involved. Three types of analysis are commonly performed by forensic laboratories: accelerant identification, comparison of evidence samples, and trace evidence examination: Each of these involves specialized analytic techniques. In addition to these techniques, photographic evidence is also a valuable tool for understanding and documenting the fire scene.

FIRE SCENE SAMPLING

Identifying what to collect and where to collect it is the most important decision an investigator must make with regard to forensic analysis of evidence. Collecting insufficient evidence or sampling in the wrong location can severely cripple the development of a complete fire origin and cause scenario. Moreover, if evidence is not collected during the initial investigation, it may be destroyed, diluted, or altered to the extent that it becomes unavailable, inconclusive, or inadmissible in court.

Accelerant Residue Samples

Considerable discussion has taken place in the fire investigation literature concerning how many accelerant residue samples should be collected at any given fire scene. Most laboratories collect and analyze statistics about the work they perform. One laboratory has suggested that a specific number of samples is adequate for achieving positive results. Collecting more samples that the specified number, they argue; will produce only negligible results. However, the ratio of positive samples to negative samples submitted by any given fire investigator may prove little or nothing about the investigator's sampling techniques. In fact, good

investigation technique should include submitting samples to confirm and refute the presence of accelerant residue. In practice, only the experienced fire investigator can determine how many samples will be required to adequately document a particular fire. Statistical evidence is important in ensuring quality control, but it should be interpreted with caution..

The presence of flammable or combustible liquid residue, particularly where such evidence would not normally be expected, may be a good indicator of incendiarism. However, many other cases involve the collection and analysis of such debris samples to rule-out the presence of accelerants or to identify possible flammable or combustible liquid residues from sources other than accelerants.

The following general principles extracted from the fire investigation literature should apply to the selection of debris samples:

- The collection of debris samples for identification of accelerant residue should begin as, soon as possible after fire control efforts have been completed. 'Fire effects, evaporation, and dilution or dispersion by fire streams all diminish the chances of collecting a positive sample where one should be present.

- The best place to sample for accelerants is at the periphery of so-called pour patterns. This is often evidenced by a distinct interface between the most heavily damaged area and the less damaged areas.

- Absorbent materials like carpet and padding, soft woods, fabric from clothes, linens, or drapes, paper,, and soil make better samples than non-absorbent, materials like glass, tile, and concrete.* However, some types of concrete actually, make good samples, especially when the sample is particularly porous and collected before the potential accelerant sample can be diluted, dispersed, or evaporated.

- Clean, dry absorbent cotton, sanitary napkins, or clean, dry cotton baby diapers make good absorbent materials for collecting flammable or combustible liquid residues on non-porous materials. Commercially available absorbent pads, such as those made of polypropylene, are unsuitable for, this purpose because they often contain background contamination: These products are commonly used in hazardous materials clean-up and are designed specifically for that purpose, not evidence collection.

- Calcium carbonate or activated charcoal may be used to adsorb accelerant residues from concrete. A thin layer of one of these materials spread evenly over the surface where a residue is suspected can collect a significant quantity of liquid. The material and a small sample of the non-porous surface should be collected and submitted for analysis.[3]

The container in which the accelerant was transported to or used at the scene is the best source of an accelerant residue sample.[4] Even fire damaged containers often contain sufficient flammable or combustible liquid residue to permit analysis. Moreover, container labels and markings often provide additional valuable clues.

Trace Evidence

Trace evidence includes a very wide array of materials or clues other than accelerant samples which may indicate the identity of a suspect(s), the motive(s), or the origin and cause of the fire. Items under this heading include:

- *Serological samples,* e.g., blood, hair, and skin, from which blood type or a DNA "fingerprint" can be identified. This type of evidence is often indispensable in identifying or eliminating possible arson suspects. As the scientific methods associated with DNA fingerprinting become more widely accepted and the procedure more common, this may also become a useful way of identifying fire victims.

- *Questioned documents,* e.g., financial records, checks, letters, business papers, insurance policies or certificates, securities, and property records or titles. When, why, and how a document was prepared as well as what information it contains may provide valuable clues about why a fire happened and who may be responsible for setting it if arson is suspected.

- *Latent fingerprints.* Fingerprints can be a valuable means of identifying suspects in arson cases, as well as establishing positive identification of fire victims. Latent fingerprints may often be recovered from other forms of trace evidence, such as Molotov cocktails, accelerant containers, and failed incendiary devices. Even

small fragments of these containers or devices may contain sufficient information to permit a positive identification.

- *Device remnants,* e.g., pipe parts or fragments, machine or appliance parts, wire or metal samples, container fragments, chemical incendiary residues, and other materials. Many forensic science laboratories maintain extensive exemplar files of materials used in the manufacture of incendiary and explosive devices. Even small fragments can often be identified using undamaged original pieces from these files. Many other articles are extensively catalogued and can be used to identify remnants of virtually any device or component. Appliance parts can often be identified using repair, guides and maintenance manuals. Wiring and metal, parts often provide valuable clues about localized heating or temperatures at different points in the fire if examined closely by a trained technician. Chemical incendiaries may be manufactured from a wide variety of common consumer products such as swimming pool disinfectant, hair gel, soap powder, brake fluid, and road flares. Often residues from failed incendiaries provide valuable clues about the ignition scenario.

- *Toolmarks.* Most tools used for prying, lifting, and cutting leave distinctive marks. These marks can be used to identify the source of damage to doors, windows, cabinets, and other secure locations. Often this is important when ruling out that damage was caused by firefighters forcing entry for firefighting. Similarly, marks on bodies can be identified in a similar fashion. Exemplar tools and photographs are used to establish similarities between the geometry of the mark and individual tools. Comparisons with tools or devices collected at the scene can sometimes establish whether a specific tool or device was responsible for the damage.

The kinds of trace evidence collected will be most dependent upon the nature of the fire and the types of evidence available. Sometimes trace evidence is disregarded in favor of collecting accelerant residue samples. This may be a mistake, especially when the presence of accelerants is uncertain or the fire scenario is unclear. The best trace evidence is often that for which an explanation is not apparent. For instance, finding melted metal parts near a point of origin may be curious but totally inexplicable at the time they are collected. However, later analysis may later not only provide an explanation but suggest a whole new investigative approach.

A main problem in collecting trace evidence is the small quantity of valuable trace evidence one would expect to find in a considerable volume of useless fire debris. Determining what will provide information of value may be quite difficult due to the extent of damage and the quantity of debris present. Destructive overhaul practices may also destroy valuable trace evidence and produce excess debris. Sieving debris in the field is one approach to this difficult problem. However, sieves may damage some brittle specimens, rendering them worthless. Although time consuming, the best approach is simply a thorough and systematic search of the scene.

Unlike other material evidence which may simply be picked up or extracted and placed in a container, fingerprints require highly specialized collection techniques. In most cases where fingerprint identification is necessary, either special care must be taken to preserve the sample or the fingerprints must be "lifted" in the field by a trained technician. For example, if an unbroken or defective Molotov cocktail is found at the scene, usually the investigator will want to identify the contents to determine if it is an accelerant and will want to identify the perpetrator(s) who built and/or used the device. Preserving the sample for accelerant analysis, however, will destroy the fingerprints on the outside of the container. Therefore, either the contents must be decanted into an evidence container for separate analysis and the Molotov container preserved for latent fingerprint analysis, or the prints must be lifted in the field and the entire device shipped to the lab for analysis. Most investigators prefer the former method.

Comparison Samples

Comparison samples are materials or objects which are believed to be nearly identical to similar accelerant debris samples with the exception that they are not believed to contain accelerant residues. The purpose of such samples is to identify and minimize or eliminate sources of interference in the analysis of such samples. Examples of comparison samples include the following:

- *New, unused evidence containers.* Most evidence containers yield no chromatographic interference. However, some types of containers, like certain types of plastic bags and specially coated containers, may give off background vapors which may obscure chromatographic analysis of evidence. New types of containers and new lots of certain container types, like plastic bags, should be periodically submitted to the laboratory for chromatographic analysis as comparison samples. Likewise, collection materials may sometimes present a source of interference. For example, some absorbent materials for

controlling petroleum spills are unsuitable for evidence collection because they have been found to contain traces of contaminants.

Comparable accelerant samples. Often investigators need to identify similarities between samples of known flammable and combustible liquids collected at the scene, from those found in containers or obtained from other sources near the fire scene, and flammable and combustible liquid residues found in fire debris. Flammable and combustible liquid residues in debris samples may come from sources inherent to the scene, or may be brought to the scene from another source. For example, insecticides often contain a petroleum-based carrier. Comparing debris samples with samples of liquids from these sources and with comparison samples of materials similar to those present in the fire debris may be a valuable method of distinguishing between what was at the scene before the fire and what was brought to the scene to start or spread the fire.

Comparisons are often requested by fire investigators when statements, evidence, or circumstances indicate that an accelerant may have come from a specific source, like a suspect's automobile or a nearby gas station. Forensic science analysts can occasionally discriminate between samples of similar materials as having clearly different sources. However, if no differences are found, the strongest statement which can be made is that the samples may have had a common source. This is due largely to the marketing practices of the petroleum industry and the near impossibility of accounting for every possible source. Another possible conclusion is that the samples belong to the same or different classes of petroleum products identifiable using gas chromatographic pattern recognition. The best samples for such comparisons are liquid-to-liquid samples. It is much more difficult to determine a common source using liquid samples and debris samples.

Comparable material samples. It is desirable to collect samples of materials similar to those found in the accelerant debris matrix that are identical except that they contain no accelerant. These materials, such as carpet or wood trim, may often give off pyrolysis products which partially obscure accelerant patterns during analysis. A comparison sample should be collected for every questioned fire debris matrix.[5] The same can be true of generic collection materials such as disposable diapers.

Comparison samples can be very important to making a positive identification of a material. Whenever there is any doubt about whether a comparison sample is needed, one should be collected and submitted or the laboratory should be contacted.

EVIDENCE PACKAGING

Evidence packaging is important for preserving and protecting evidence between the time it is collected and when it can be analyzed in the laboratory. In some cases, it may take several weeks to have evidence processed by a forensic science laboratory. The nationwide war on drugs has significantly increased the turnaround time for evidence at most state and regional forensic science laboratories due to the heavy drug-related caseload. Therefore, fire investigators should take particular care in packaging and preserving evidence to maintain the quality of the sample, and prevent release, deterioration, or contamination.

Accelerant Residue Samples

Because of the volatile nature of flammable and combustible liquids, debris samples must be placed in vapor-tight containers. According to laboratory supervisors, the best container for flammable or combustible liquid accelerant residue and debris samples is unused metal paint cans with tight-fitting friction lids. Laboratory personnel cite the advantages of the containers' large openings in relation to overall container diameter, vapor-tightness, rigid form, lightweight, durability, lack of appreciable background interference during analysis, availability, and range of available sizes. Other containers have their uses as well; the container must be compatible with the sample and the tests or analyses to be performed.

Samples such as clothing or textiles may be too large for metal cans. In such cases, many fire investigators have found Kapak™ bags to be a good substitute. These self-sealing bags have an exterior layer of polyester and an interior layer of polyethylene. Many other types of plastic bags and containers may not be well-suited to use as accelerant residue debris containers. Polyethylene plastic bags, paperbags, aluminum foil, mylar bags, and coffee cans with plastic lids are not vapor-tight with respect to flammable and combustible liquids. Petroleum products will pass through paper and polyethylene, and be lost to the environment or contaminate other samples they are packaged with. Other types of plastic bags, such as Kapak™ or nylon, may tear if the debris sample contains sharp objects such a glass or nails.

Liquid samples may be placed in small glass vials with screw-top lids, not rubber stoppers. These containers should then be placed inside a metal can filled with Vermiculite™ to prevent the vials from breaking in transit.

Some fire investigators prefer glass Mason™ jars for collecting and preserving evidence. The primary advantages are the availability of these containers at hardware, variety, and farm stores, especially in remote or rural areas away from central cities. With proper protection and packaging these containers can be excellent for collecting and preserving accelerant residue samples. The screw-top vapor-tight lids work well and the wide mouth makes it easy to insert and remove samples. Moreover, the sample remains visible, permitting inspection without opening the container. However, some petroleum products and volatiles present in fire debris may degrade the jar lid's rubber seal, allowing vapors to escape. One proposed solution to this problem is the insertion of a layer of aluminum foil between the jar opening and the lid.

The quantity of material needed to perform a successful analysis is often misunderstood. Only a relatively small quantity of a very concentrated sample is needed. The more dilute or evaporated a sample, is the more material will be required to perform a good analysis. Although small samples may be adequate under the right circumstances, in general it is better to collect a larger as opposed to smaller sample. After all, it is often difficult to tell precisely where in a given sample area the greatest concentration of accelerant will be found. Samples should be collected at each location where accelerant residue is suspected. A good sized sample will fill a one quart or one gallon metal can one-half to three-quarters full. Carpet samples should be about as wide as the can is deep, and as long as the can is round. A good carpet sample will fully line the interior of the can. When residue is extracted from concrete, all of the absorbent material and a small amount of concrete from the surface area where the adsorption was performed should be collected and packaged together. If a liquid is wiped or absorbed from another non-porous surface using a clean, dry absorbent textile material, as much residue as practical should be collected and the entire absorbent material submitted.

Trace Evidence

Trace evidence covers many different types of specimens and samples, and therefore, packaging methods vary widely as well. The packaging technique used must be compatible with the type of evidence collected and the test(s) to be conducted. Many trace evidence specimens will be analyzed by more than one technique.

Serological specimens and samples for latent fingerprint examination should be sealed in brown paper bags or carefully wrapped in kraft paper. Air- or vapor-tight containers will cause deterioration of such samples due to bacterial growth of serological specimens. Latent fingerprints will be destroyed by moisture or petroleum products. It is better to have the sample exposed to air than dissolved by moisture or contaminated by bacteria growth.

Documents can provide valuable clues not only about what happened but why it happened too. Questioned documents should be packaged in clear, protective envelopes to avoid wear and tear and protect latent fingerprints. Partially burned or water-soaked documents should be placed on a firm support, such as cardboard, poster board, dust pan, or cake plate, and placed in a flat box for transport to the laboratory.[6]

All types of trace evidence, especially device remnants and unidentified debris, should be carefully described by the investigator. Extra care should be taken to give the laboratory analyst any clues the fire investigator may have about the article's identity without presupposing any particular conclusion. Likewise, the investigator should give the analyst a clear idea of what he or she wants to know about the item, e.g., latent fingerprints, blood type, physical match with comparison sample, and so on.

Comparison Samples

Comparison samples should be packaged exactly the way similar accelerant debris or trace evidence samples are packaged. Comparison samples are submitted for the purpose of minimizing or eliminating interference from sources other than an accelerant or identifying similarities between "knowns" and "unknowns." To ensure the greatest degree of reliability, the practices used for processing and preserving comparison samples must be as similar to those used for the target sample as practical.

Storage and Preservation of Evidence

With the delays in processing evidence growing in many places around the country, proper storage is increasingly important. Some agencies may find themselves responsible for evidence samples due to delays in shipment, laboratory backlogs, or samples returned from one laboratory awaiting shipment to another laboratory. Under such circumstances, fire investigators should be prepared to store, document, and preserve the evidence in the same manner as it would be while awaiting analysis at the laboratory. American Society for Testing and Materials standard ASTM E1492-92, *Practice for Receiving, Documenting, Storing and Retrieving Evidence in a Forensic Science Laboratory* details the specific practices recognized throughout the forensic science community for maintaining the chain-of-custody and preserving evidence samples before and after forensic analysis.

FORENSIC TECHNIQUES

Modern forensic science provides valuable tools for documenting and verifying the findings of field investigators. Forensic science applies the principles of analytical chemistry and physics to the identification, classification, and comparison of evidence samples collected by fire investigators for the purposes of determining the truth about fires and their causes. These techniques often involve complex, specialized equipment, but more importantly, they require the competence and professionalism of dedicated field investigators and laboratory technicians working together to obtain and analyze the evidence.

Accelerant Identification

Most flammable and combustible liquids used to accelerate the spread of a fire are products refined from crude oil. Crude oil and its refined products, such as gasoline, kerosene, and diesel fuel, are complex mixtures of thousands of different organic compounds. These compounds are mostly hydrocarbons; that is, compounds containing only carbon and hydrogen atoms. Gas chromatography is an analytical laboratory technique used to separate mixtures of volatile organic compounds, and therefore, is well suited for analyzing petroleum products recovered from fire debris samples.

Accelerant identification in the laboratory consists of three steps:

- *Sample preparation,* the process by which the liquid is extracted or isolated from fire debris.

- *Instrumental analysis.* Once isolated, a sample is analyzed using gas chromatography (GC). GC separates components of a mixture by volatilizing the sample into a carrier gas stream, which transports it through a column containing a fixed absorbent phase. Upon emerging from the column the fractions are ionized and exposed to a detector which registers the relative concentrations to permit analysis. The concentration results are plotted on graph paper as a function of time. This graph, called a chromatogram, is interpreted by an analyst to classify the material. GC is a good technique for analyzing extremely small quantities of complex mixtures.

- *Data analysis.* The determination whether a petroleum product is present in the sample, and if so what the product is, is performed by comparing the sample chromatogram with other chromatograms of known products or comparison samples to produce an identification.

Sample Preparation

The isolation and recovery of a sample is the first and most important step in identifying a possible accelerant residue using gas chromatography. There are five different methods for recovering petroleum products from fire debris:

Direct headspace - sampling from the vapor space in the container above the debris sample which may either be heated or at room temperature.

Static head space - placing a material such as activated charcoal in the debris which will adsorb the volatile compounds present.

Dynamic head space - sweeping air into the container then extracting it over some material such as activated charcoal which will adsorb the volatile compounds present in the vapor stream.

Solvent extraction - placing a solvent such as carbon disulfide (CS_2) in the container.

Steam distillation - no longer used because it the least sensitive and most time consuming.

Static and dynamic headspace techniques are among the most commonly used since they are very sensitive and require little labor. American Society of Testing and Materials standards have been adopted to document each of these procedures.

·ASTM E1385-90, *Standard Practice for Separation and Concentration of Flammable and Combustible Liquid Residues from Fire Debris Samples by Steam Distillation.*

·ASTM E1386-90, *Standard Practice for Separation and Concentration of Flammable and Combustible Liquid Residues from Fire Debris Samples by Solvent Extraction.*

·ASTM E1388-90, *Standard Practice for Sampling of Headspace Vapors from Fire Debris Samples.*

·ASTM E1412-91, *Standard Practice for the Separation and Concentration of Flammable and Combustible Liquid Residues from Fire Debris Samples by Dynamic Headspace Concentration.*

·ASTM E1413-91, *Standard Practice for Separation and Concentration of Flammable or Combustible Liquid Residues from Fire Debris Samples by Passive Headspace Concentration.*

Instrumental Analysis

National standards for forensic analysis of fire debris have recently been adopted which recognize GC as the *only* acceptable means of identifying flammable or combustible accelerant residues. In recent years, improvements have

---

Table 4.1

Comparison of Gas Chromatographic Techniques

Advantages/Strengths	Disadvantages/Weaknesses
Capillary column gas chromatography offers best resolution over shortest time of current analytic techniques	Has difficulty identifying highly volatile substances and polar solvents due to lack of generally acceptable enrichment techniques
When used with mass spectormetry, interference can easily be filtered out permitting positive class specific identification	Mass spectrometry adds to the time and cost associated with processing samples, and is therefore, seldom used.
When used with data analysis routines (pattern recognition programs), operator bias can be reduced, cost-effeciency improved, and statistical control achieved.	Without data interpretation, analysis is still carried out by visual interpretation and comparison

Source: Bertsch, W., Zhang, Q.W., and Holzer, G. (Sept. 1990), "Using the the tools of chromatography, mass spectormetry, and automated data processing in the detection of arson, *"Journal of High Resolution Chromatography,* Vol. 13, No. 9, p. 601.

Table 4.2

Accelerant Classification Scheme		
Accelerant Class and Category	**Approximate Boiling Point Range [°C]**	**E x a m p l e s**
1 Light Petroleum Distillates (LPD)	<120	Petroleum ethers, pocket lighter fuels, rubber cement solvents, skelly solvents, lacquer thinners, VM & P Natptha
2 Gasolines	50-220	All brands and grades of automotive gasoline - including gasohol, some lantern fuels
3 Medium Petroleum Distillates (MPD)	60-200	Charcoal starters, paint thinners (Oil based), mineral spirits, "drycleaning" solvents, torch fuel
4 Kerosene	90-290	No. 1 fuel oil, jet-A (aviation) fuel, inset sprays, charcoal starters
5 Heavy Petroleum Distillates (HPD)	210-410.	No. 2 fuel oil, diesel fuel
0 Unclassified	Variable	Single compounds such as alcohols, acetone, or toluene, and xylenes, isoparaffinic mixtures, some lamp oils, camping fuels, lacquer thinners, duplicat. ing fluids, and others.

been made to this procedure to reduce interference from background sources within the fire debris matrix and improve reliability. These include the addition of mass spectrometry and pattern recognition computer software.

To understand the application of gas chromatography to fire investigation, and the impact of these improvements, it is important to understand a little more about how a GC works.

Petroleum products contain hundreds of different organic compounds, mostly hydrocarbons. Each hydrocarbon is characterized by a specific boiling point. This property differs for compounds based on their size and chemical structure. Gas chromatography can be used to separate mixtures based on the boiling points of the components of the mixture. A petroleum product or isolated sample is injected with a syringe into a heated injection port where it is instantly vaporized to the gaseous phase. The vaporized mixture is swept into a column by an inert gas, usually helium or nitrogen. As the hydrocarbons travel down the column, they are separated according to boiling point. The more volatile compounds having lower boiling points travel down the column faster, than the high boiling point, less volatile compounds.

As each compound leaves the column, it passes by a detector. The most common detector type used in forensic gas chromatography is the flame ionization detector (FID) because it is best at detecting organic compounds and is especially sensitive to hydrocarbons. As a compound passes through a hydrogen flame in the detector, it is ionized. The presence of charged particles, ions, within an electrode gap causes a current to pass through the gap. The current is measured and amplified by an electrometer, which sends a signal to a recording device, such as a mechanical chart strip recorder or a computer. The resulting chart is known as a chromatogram. This chart contains many peaks, each of which correspond to at least one chemical compound. Collectively the peaks form a pattern. Patterns of chromatograms of samples isolated from fire debris are visually compared to patterns of chromatograms of known petroleum products to obtain an identification.

To' ensure reliability, repeatability, and consistency, the forensic science community has developed guidelines for conducting laboratory analyses of fire debris. These procedures are documented in ASTM E1387-90, *Standard Test Method for Analysis of Flammable or Combustible Liquid Residues in Extracts from Samples of Fire Debris by Gas Chromatography.* Another standard, ASTM E1389-90 *Standard Practice for Cleanup of Fire Debris Sample' Extracts by Acid Stripping,* covers one cleanup method to remove interfering compounds arising from the sample matrix.

Data Analysis

Most chromatograms are still visually analyzed or "eye-balled" to 'identify' the material present in the sample. Although each petroleum product produces a distinctive chromatogram based on its chemical composition, background volatiles found in most fire debris may cause even similar fire debris samples to produce different chromatograms.

Pattern recognition is the method used to identify petroleum products recovered from fire debris. Forensic laboratories have examined hundreds of commercially available petroleum products and found their chromatographic patterns fall into the five categories shown in Table 4.2. Because of the blending processes used to manufacture automotive gasoline, it produces a pattern unique enough to be in a class of its own. The other classes of products represent "cuts" off the distillation tower at the refinery. Note that these classes contain a variety of commercial end use products. It is not possible for the forensic science analyst to determine what the exact commercial end use was of a product detected in a fire debris sample,. For example, the same medium petroleum distillate (MPD) may, be sold commercially as either paint thinner, mineral spirits, a dry cleaning solvent, or a charcoal lighter fluid. The fire investigator must keep in mind that a petroleum product detected in fire debris may come from a variety of sources, some of which are inherent to the scene.

The sample preparation techniques used to recover petroleum products from the fire debris also recover combustion and pyrolysis products from the sample matrix. The sample matrix may consist of wood, carpet, or other furnishing and building materials in addition to flammable or combustible liquid-residue. The presence of these materials may complicate the chromatogram to a point where the identification of a petroleum product is not possible due to interference from pyrolysis products. In some cases, using a more selective and specific detector can help. One detector used for this purpose is the mass spectrometer. Where the flame ionization detector merely detects the presence of organic compounds, the mass spectrometer can determine what specific chemical compound is coming out the end of the GC column. Using various computer data manipulation routines, gas chromatography/mass spectrometry (GC/MS) can sometimes identify petroleum products in samples containing interfering peaks in the GC-FID chromatogram.

GC/MS is also useful for identifying products 'that are unusual or contain only a few chemical compounds and therefore do not produce a distinctive pattern. One example is alcohol. Grain alcohol (ethanol), wood alcohol (methanol), and rubbing alcohol (propanol or iso-propanol) all consist of a single chemical compound, and will produce only one peak on a GC-FID chromatogram. GC/MS on the other hand, can identify that the compound is indeed, for example, ethanol. Being volatile and water soluble, alcohols are not likely to be found in fire debris samples. If an investigator thinks alcohol may have been used to set a fire, he or she should alert the laboratory analyst so the proper steps can be taken to recover and identify the material in debris samples.

Computer pattern analysis provides another useful way of identifying accelerants from complex chromatograms. Computers and specialized software employing complex statistical techniques may be used to compare the sample chromatogram with othersstored in the computer's data files and analyze the correlation between them. This technique is limited, however, by the complex software required to perform the calculations and the fact that mini- or mainframe computers are required to perform the calculations. To date, computer pattern recognition has been applied to identification of the source of oil spills, but has not been widely adopted for use in the analysis of data from fire debris samples.'

After the chromatogram has been generated, it is interpreted. Five categories are used to classify flammable and combustible liquids. These are illustrated in Table 4.2. Fire investigators often request that forensic science laboratories identify a specific product such as gasoline by brand or grade. The nature of gasoline refining and marketing currently make this impossible. It is not practical to describe a sample in terms more distinctive than these fivecategories: light petroleum distillates, gasoline, medium petroleum distillates, kerosene, and heavy petroleum distillates.

Examination of Trace Evidence

Examination of trace evidence is conducted either to identify the material or its source, or associate the item or specimen with another sample. A vast array of techniques is used to analyze trace evidence. Most of these techniques involve the careful, detailed examination of the article or specimen. Special equipment may be used to enhance the analyst's visual capabilities and produce permanent records or images of the observations made. The most popular instruments for analyzing trace evidence are the magnifying glass and stereo microscope. Scanning electron microscopes, metallurgical, x-ray diffraction, infrared, ultraviolet, and other techniques are also frequently employed.[8]

An example of the power of trace evidence examination is illustrated in the techniques employed in the investigation of a bombing. Pipe fragments from a device may be recovered during on-scene investigation and shipped to the laboratory for examination. During examination, the analyst carefully and systematically searches the surface for chemical residue. If a particle of smokeless powder is discovered, its shape, color, and texture are classified and compared with the characteristics of smokeless powder samples on file with the laboratory to make an identification. The pipe fragments themselves are compared with exemplars or samples of undamaged pipe from a wide variety of manufacturers also' kept on file in the laboratory. Finally, the fragment is examined for the presence of latent fingerprints. Any prints found are lifted and classified. The fingerprint classification is then compared with those on file in a national database and if a "hit" is obtained, the actual fingerprint samples are compared.

Other types of evidence undergo similar examination processes. For example, an analysis of a questioned document may be conducted to discern what information it contains and determine its origin. Likewise, latent fingerprints may be lifted from such documents. Forensic specialists known as "questioned documents examiner" analyze the origin and preparation of documents. Latent fingerprint examiners identify, classify, and analyze fingerprint evidence. To ensure that both procedures produce the proper results, the document should be examined by the questioned documents examiner first. This way any damage caused during the latent fingerprint examination will not obscure or destroy the evidence of the document's origin or method of preparation or alteration. Serological samples are chemically analyzed to identify blood type, perform DNA fingerprinting, or determine specimen morphology for comparison with samples from suspects.

Comparison of Evidence Samples

Comparison samples are processed in the same manner as the counterparts to which they will be compared. The collection and analytic techniques should be as similar as possible to ensure validity and reliability.

PHOTOGRAPHIC PROCESSING AND ANALYSIS

Photographic documentation of the fire scene is perhaps the most common of all analytic techniques. The adage that "a single picture is worth a thousand words," is certainly true when documenting or analyzing a fire scene.

This section discusses the types of equipment and supplies used to document fire scenes visually, and the application of photographic evidence is considered. For a detailed discussion of how to photograph the fire scene, see the excellent references on this subject in Chapter 6.

Fire Scene Photography

Without going into detail about how to take pictures of the fire scene, it is important to understand what to photograph.

Photographs should serve as a visual record of the circumstances surrounding the origin and cause of the fire. The photographic record should, to the extent practical, present a visual report of the fire.

Two general rules of thumb should apply to fire scene photography:

'Work from the general to the specific. Each photographic sequence tells a story. A good story always describes the setting in detail. The best composite of the fire scene story is told by weaving the details in close-ups into the larger mosaic of the fire as a whole.

'Don't be afraid to take too many pictures. Photographic film and processing are a lot cheaper than a lost opportunity to clear a case. Failing to document a key piece of evidence or document how it was discovered or processed can damage an otherwise credible case. Reason should prevail in deciding what to take pictures of or how many to take, but it is usually better to err on the side of more rather than less. Besides, there are plenty of ways to economize in the way the photographs are taken and processed rather than simply not taking them at all.

During the fire -- Fire scene photographs taken during the fire can be a valuable adjunct to the investigative record, especially if they can be tied to a particular point in time. Fire investigators can use these pictures to obtain a better understanding of the fire's development while it was in progress. Vehicle-mounted videotape cameras are another way of producing a detailed visual record of the fire while it is still in progress. Many police agencies have recently begun installing these devices in large numbers to record police activity like traffic stops. Many of these cameras also come equipped with time and data stamps.

Exterior -- Photographs of the outside of the building document what the firefighters themselves may have seen when they arrived. Smoke stains, thermal damage, and structural collapse all document what effects the fire may have had on the inside, as well as the potential effects on adjacent structures. Exterior photographs should illustrate the following key features:

- Property address or location
- Relationship to adjacent properties or structures
- Condition on each side beginning with the "front," that being the side to which the fire department first responded
- Conditions at specific locations where smoke or fire vented or collapse occurred

Figure 4.3

	Anytown Fire Department Fire Investigation Unit Photographic Log		
Address: 4321 W. Broad Street	Investigator: B. Dano	Date/Time: 1-1-91 12:01 am	
Photo No. (Roll/ Exposure)	**Subject**	**Exposure Information**	**Flash**
A/1	Exterior front (side 1) from across street, Note: Heavy fire damage above front door with smoke staining and radiating vee pattern.	35 mm Wide f2.8 - 1/60	X
A/2	Exterior east side (side 2) from front porch of adjacent dwelling. Note: Broken windows on first and second floors, smoke stains over 1st floor.	50 mm f1.7 - 1/60	X
A/3	Exterior rear (side 3) from alley. Note: No fire damage, door forced open, windows broken on first floor.	35 mm Wide f2.8 - 1/60	X
A/4	Exterior - closeup rear door from back porch. Tool marks look like FD forcible entry.	70 mm Zoom f16 - 1/60	X
A/5	Exterior west side (side 4) from side yard of adjacent property.	35 mm Wide f1.7 - 1/60	X
A/6	Exterior - front entrance door open. Taken from porch. Note: High heat damage and diagonal burn pattern.	50 mm f3.5 - 1/60	X
A/7	Exterior - closeup on front door lock and hardware from porch side.	70 mm Zoom f16 - 1/60	X

Entrances -- Each entrance to the building should be carefully examined and photographed to document its condition at the time the fire occurred and during the incident. Close-ups of tool marks will help corroborate witness accounts and may aid forensic science analysts in determining what tool(s) may have caused the damage. Locking and security hardware should be carefully photographed to document how the scene was secured and how it may have affected occupant egress.

Interior -- Photographs should document the inside beginning with the areas of least damage and working toward the area of heaviest damage. This is generally the preferred investigative sequence as well. Photographing as you go helps tell the story of how the fire was investigated and confirms that proper investigative procedures were employed throughout.

Photographic Log -- It is a good idea to keep a log of what was photographed, where it was photographed from, and how the picture was taken. This information may be captured textually, verbally, graphically, or by a combination of these means. It is often difficult to fully understand what a photograph was intended to illustrate if this information was not recorded. Even properly exposed photographs are sometimes very difficult to interpret due to the nature of the subject. Likewise, fire scenes often produce extremely poor photographic conditions. Where a photograph does not come out properly, it is often helpful to know why it did not. The photographic log will also documents the sequence the photographer used to approach the subject investigated and photographed. Since photographs appear in time sequence on a roll of film, this is a useful way of demonstrating that a systematic method was used to investigate and document the fire scene.

Figure 4.3 details a few entries from a good written photographic log. Note that it contains an identification of the photograph as it appears on the roll; a description of what is in the picture; where it was taken; what lens, aperture, and exposure were used; and, whether or not it was taken with a flash. The same information can be read into a small tape recorder or into the microphone of a video camcorder. Many fire investigators also produce detailed floor plans to document what was photographed and where the photographer stood while taking the picture.

Photographic Equipment and Techniques

Almost any camera is better than no camera. Taking good pictures is more a matter of the photographer's skill than the quality of the camera. Today more than ever, the introduction of low cost high-quality photographic equipment and high-speed film have made fire scene photography more practical than ever before, even for small departments and those on tight budgets.

Cameras, Lenses, and Accessories -- Most fire investigators seem to prefer 35-mm single lens reflex (SLR) cameras with interchangeable lenses. These cameras give the user control of aperture, film speed, shutter speed, and flash. Fire investigation units on tight budgets, fire company officers, and inspectors have found the new generation of 35-mm automatic cameras with fixed lenses quite useful. Many of these cameras come equipped with so-called data-backs which imprint the time or date on the photographic image. Some even adjustable lenses which provide a range of focal lengths from close-up to zoom. More recently, automatic settings and features have been combined with the traditional features of SLR cameras to give the photographer a full range of creative options. These cameras have made high-quality photographic capabilities accessible to almost all departments and make fire scene photography increasingly practical.

When selecting a camera for fire scene use, the fire scene environment should be a prime consideration. Dust, dirt, water, and corrosive gases and vapors may all damage the camera and accessories. Most professional SLR cameras can take this kind of abuse if they are handled with adequate care and cleaned and maintained regularly. Lenses will be one of the most vulnerable parts of the camera, and should be covered when not taking pictures. To provide protection during photographing, a clear or ultraviolet filter should be attached to prevent the lens' surface from being scratched by fire debris. It is much cheaper ($15 - $20) to replace a filter than a lens ($75 on up).

As already mentioned, fire scenes are not the best photographic environments. Fire and soot blackened surfaces make getting good contrast extremely challenging. Flash units may be particularly susceptible to water damage Flash units should provide adequate power to fully illuminate even very dark subjects.-Many camera kits come with their own flash units, but these are usually designed for "normal'? picture taking; more powerful units may be needed for fire scene work. Keeping power supplies and flash units covered when not taking pictures and avoiding water spray at all times are generally good rules of thumb.

Many cameras now come equipped with built-in automatic winders. Some photographers use external motor drives and/or automatic winders to advance and rewind the film. Like other motor driven devices, motor drives and automatic winders may be highly prone to shock and impact damage. External motor drives and automatic winders run approximately $150 to $200.

A tripod or monopod makes an excellent accessory. Large apertures and slow shutter speeds are often required due to the poor light conditions and contrast found on the fire scene. These conditions can cause pictures to blur easily. A tripod or monopod provides an ideal support from which to shoot under these difficult conditions. A good tripod costs about $100.

Again, fire scene conditions conspire against otherwise good photographers. Poor lighting and contrast will make a flash a necessity rather than an option. Under normal lighting conditions a light meter will help the photographer determine the best shutter speed and aperture to use to obtain a given exposure. Most flash units require that a single shutter speed be used in order to maintain synchronization between the shutter and the flash. A flash meter will do the same thing as a light meter except it works with a flash. Instead of reading the amount of ambient light available, a flash meter measures the amount of light reflected by the subject under flash lighting conditions and recommends the appropriate aperture to achieve proper exposure. Good flash meters start at approximately $120 and run into the neighborhood of $350.

Film -- A wider variety of film is now available. Moreover, the high speed films of today produce higher quality images. A great many options are now available to the fire investigator for documenting the fire scene. The photographic and fire investigation texts referenced in Chapter 6 include extensive discussions of film speeds and types.

Most fire investigators report that they prefer 35mm color slides for fire scene photography. Color slides have several advantages over color and black-and-white prints.

- Cost less to process
- Better image quality
- Ability to produce duplicates and prints from same slide
- Ability to project images for courtroom presentations

The single greatest drawback of slide film is that it is slightly more expensive to produce prints from slides than from negatives. Since a slide is a positive image, an internegative must be produced before a print can be made. Usually this results in only a slight loss of color and image quality. Although slides and prints can be produced from other photographic media, the duplication process can be quite complex. Moreover, with each step in the reproduction or duplication process, some loss of clarity or color contrast occurs. Slides produce the best reproduction results for the money.

Other options are available which produce better results for certain situations. For example, print film provides the photographer with wider latitude than slide film in terms of exposure. Since the final product is a photographic print, an under exposed negative can be corrected simply by exposing the print for a longer period and vice versa. Another option has become available in the past several years which yields both prints and slides. At least three mail-order vendors have begun offering dual media films which produce image quality comparable to slides while providing the flexibility to produce a print without having to create an internegative.

Instant photography has gone from fad to mainstream. Today, instant photography has become a popular tool for capturing images quickly when only a single copy is required. The drawbacks of most instant photography are poor depth of field, poor image quality, poor adaptability (most of the cameras only have a single, fixed lens),, and film expense. However, some evidentiary purposes lend themselves particularly well to this technique. The most popular manufacturer of instant cameras and films has developed products aimed specifically at capturing macro images of small objects. Moreover, they have developed services for duplicating their photographs when the need does arise. Generally this involves taking a picture of your picture under very controlled conditions. Nonetheless, instant photography remains one of the less popular fire scene photographic techniques.

> *Video Camcorders* -- Video camcorders have become one of the most popular items on fire investigation unit wish-lists. Videotape offers several advantages over traditional still photographs. Perhaps the most important of these is the ability to capture the whole fire scene in context. Each videotape can act as a guided tour of the fire scene. As lay people become more and more familiar with videotape in their own homes, they may be expected to become more impressed by it as jurors when it is used to document the fire scene.

Another of videotape's advantages may also be a weakness. Most videotape camcorders offer the ability to record sound as well as visual information. As such the accounts and descriptions of the investigator may be captured as the scene is processed. Consequently, fire investigators using videotape should be very careful about what they say while the camera is "rolling." Preliminary conclusions and statements about suspects or suspicions about probable cause could become part of the record. Videotape used in the preparation of an arson case can be subject to the rules of discovery, and these statements can then be used to attack the investigator's credibility and independence.

The ability to edit and enhance videotape is another advantage with potential drawbacks. Editing videotape gives the investigator the ability to "cut to the chase," and bypass unimportant details so long as the complete unedited version remains accessible to the defense. Titling and captioning videotape gives the fire investigator the ability to reinforce or clarify key points about the information on the videotape and can be powerful tools in the use of videotape in the courtroom. However, the videographer must be ready to demonstrate that the images presented, after editing, titling, or captioning, remains an accurate and factual representation of the fire scene or subject.

Additional advantages of videotape are its ease of duplication and case of presentation. Several copies can be made from the same original. With good duplication facilities the number of copies from a single original is virtually limitless. Also, an almost unlimited number of people can view the same videotape, either individually or simultaneously depending on the sophistication of the broadcast capabilities.

The two most popular formats currently are ½ inch VHS and 8-mm. Whichever format is selected, the final product should be compatible with the intended playback equipment or duplication facilities should be available. The expense of videotape as a fire scene tool ranges from a little less than $1000 for a good camcorder to several thousand dollars for advanced editing, titling, captioning, and broadcast capabilities.

Using Fire Scene Photographs

Fire scene photographs are generally taken to document the conditions found during the investigation so that they can be preserved after the scene is destroyed, altered, or restored. Their principal use is in presenting a visual account of the fire and its effects. Recently a serious debate has emerged within the fire investigation community about whether or not fire scene photographs can be used in lieu of on-scene investigation to evaluate certain conditions related to a fire.[9] Those who support analysis of photographs in lieu of on-scene investigation readily admit that it is a poor substitute for on-scene investigation. It is sometimes difficult to tell whether a photographic detail comes from the photographic process or was something real on the fire scene. The value of photographs is obviously limited to that which they specifically portray. In many cases, this information, out of context, is meaningless.

Nevertheless, the use of photographs to analyze events after the fire scene investigation is complete is widely recognized as a legitimate investigative technique. In forensic analysis, photographs are often the only record of the position or condition of an object before it was collected or recovered for analysis. Consequently, the analysis of photographs should never be dismissed out-of-hand. However, photo-analysis should never replace thorough on-scene investigative efforts either.

Fire scene photographs and visual information can be presented a number of different ways with dramatic appeal and effect. The format and presentation chosen should try to satisfy each of the following conditions:

- Accuratelyrepresent the conditions as they exist in the field when the image is captured
- Accurately portray the relationship between the subject and its environment
- Produce adequate contrast and detail to permit all important features to be clearly discerned
- Clearly depict the subject so it can be easily understood by those who must use it

The last point is the easiest to overlook. Fire scene photographers often view themselves as the primary audience for their pictures. However, if videotape or photographs are taken for evidentiary purposes, many other viewers must also use the images. A forensic science technician may make a positive identification of an object which is crucial to the case on a good photograph. Civil trials may develop months or years after accidental fires are investigated. An attorney may decide how to proceed with a defense based on the photographic record. And a jury may decide the guilt or innocence of a defendant on the strength of photographic evidence. Even non-evidentiary photographs are often viewed by many people. Keeping the subject and the viewers in mind will make a better photo or video image regardless of the equipment used.

ENDNOTES

1. Sanderson, J.L., Ballict, C. A. and Balliet, M. A. (March 1990) "Sampling Techniques for Accelerant Residue Analysis," *Fire and Arson Investigator,* Vol. 40, No. 3, p. 36.

2. Dietz, W. E. (June 1991). "Physical evidence of arson: Its recognition, collection, and packaging, *"Fire and Arson Investigator,* Vol. 41, No. 4, p.35.

3. Ibid., p. 35.

4. Ibid., p. 34.

5. Ibid., p. 36.

6. Ibid., p. 39.

7: Ibid.

8. Custer, R. L. P. (1991), "Fire Loss Investigation, "Section 10/Chapter 1, *Fire Protection Handbook,* 17th ed., Quincy, MA: National Fire Protection Association, p. 10-7.

9. Smith, D. M. (Sept. 1990), "Investigation of fire through photographs! Impossible?", *Fire and Arson Invest-*tigator, Vol.41, No. 1,(pp. 59-60). Wagner, R. W. (March 1991) "The investigation of fire through photographs! Yes, impossible!", *Fire and Arson Investigator,* Vol. 41, No. 3, (pp. 58-59) and Sanders, D., ibid., (p.60).

Chapter 5 - COMPUTER SYSTEMS

Many of the public sector pioneers of information processing used to rely on mainframe and minicomputers for their information processing needs. While they still have their place, the computing power that once required a dedicated mainframe and a staff of programmers is accessible to anyone who can afford a personal computer and learn to use available software packages.

Personal computers have gained widespread acceptance in the fire investigation for managing data, writing reports, preparing diagrams of fire scenes, analyzing numerical data like financial statements and fire statistics, preparing graphics, and communicating with remote computers to share information or conduct research. As the number of agencies using computers has grown, so have the number and type of applications to support their needs. This section discusses specialized applications for fire investigation, as well as the basics of off-the-shelf software that may be useful to fire investigators.

HARDWARE

Before discussing the software that makes a computer operate, it is necessary to have a basic understanding of what constitutes a computer system. Regardless of the type(make or model) of computer, every system consist of five main components:

- Central processing unit (CPU)
- Random access memory (RAM)
- Storage device(s), e.g., disk drive, hard disk drive, floppy disk drive, optical disk drive, tape drive
- Input device(s), e.g., keyboard, mouse or other pointing device, text or graphics scanner
- Output device(s), i.e. video display terminal (VDT), printer, plotter, modem

The make, model, operating system, memory, display, and output devices (peripherals) must all be compatible with one another to operate properly. And all of these variables have some impact on what software your computer will run.

Ideally, the decision about what hardware to use should come after identifying and, defining the task or tasks for which the computer system will be used. The use of the computer will in turn establish what kind of software will be required to perform that task. By defining criteria for performance of these tasks, such as speed, ease of use, and price, decisions can be made about the system to purchase - including both hardware and software.

Many agencies have already made the leap to computers. In these casts, care must be taken in selecting new software to ensure compatibility with the system hardware already in use. As new software comes on the market, many users find it necessary to upgrade the memory, display, or storage capabilities of their systems to take advantage of new software features.

APPLICATION PROGRAMS

Computers are far from being "thinking machines." Actually, computers themselves are quite dumb, nothing more than a collection of inanimate parts. If computers are so dumb, how do they do such powerful things? The answer lies in the software which brings the inanimate parts to "life." Essentially, someone else has had to do the thinking. Application programs are the answer. Unless you are a computer programmer, or have the budget to hire one, you probably use application programs for everything your computer does. Most application programs for personal computers fall into one of five major categories:

- Word processing
- Database management
- Spreadsheets
- Graphics
- Communications

As this list illustrates, these programs provide you with the tools to perform a wide variety of tasks. Most fire investigators will probably find the existing application programs quite capable of servicing their computing and dataprocessing needs. Word processing, database management, number crunching, drawing and graphics, and electronic communications are no different for fire investigation than the same tasks in other disciplines.

Off-the-shelf application programs can be useful tools of the fire investigation trade. Like other tools, they are only as effective as the person using them. Application programs only instruct the computer to perform those tasks you request. Therefore, training and practice are required. Although you do not need to be "computer literate" to use the most current application programs, most fire investigators will find that a modest investment of time is required to become proficient. The emphasis on "user friendly" programs for all applications has produced programs which most fire investigators will find useful and easy to learn and use in a minimum of time. Moreover, the overwhelming majority of fire investigation computing can be performed using the off-the-shelf application programs. For those who still find the personal computing intimidating, low cost short-courses, seminars, self-instruction videocassettes, and tutorials are available.

Word Processing

Word processing or writing programs allow you to interact with your computer to product written records and documents. Modern word processing and desktop publishing (DTP) programs incorporate features formerly restricted to typesetting machines. The major advantages of word processing programs over typewriters and handwritten documents are editing, appearance, and space. Word processing software permits the writer to edit and modify text before printing. Revisions can be incorporated without producing a whole new document. This alone is a major labor-saving advantage. Advanced editing and formatting features, now part of most programs, make it easier to produce higher quality documents and publications with greater "eye-appeal." Most programs now incorporate features for viewing the document as it will appear on the printed page without wasting a sheet of paper. For the investigator, the appearance of a document can be almost as important as the content when it is used to influence or convince a prosecutor or jury. Advanced formatting and typesetting features and the use of graphics in documents make it easier to demonstrate the logic of arguments and reinforce the writer's credibility. Finally, computer generated records take up far less space than paper documents. One floppy diskette can often store hundreds of pages of information.

The disadvantages of word processing software are relatively few, the most commonly cited ones being the difficulty learning how to use a particular system, the cost of the program, and cross-compatibility between different platforms and applications. In general, it is both possible and relatively easy to minimize any one or more of these disadvantages through informed shopping. The key to finding the right software at the right price is knowing what you need to do, what you want to do, and what you are willing to pay to get the job done. The more you are willing to spend the more you will be capable of doing. Most word processing programs retail between $250 and $500. Desktop publishing programs anchor the upper end of the price spectrum, weighing-in at about $900. Concerns about the user-friendliness of a program and the quality of customer support can usually be addressed by interviewing other users or a quick survey of the popular computer-oriented media.

Database Management

A database is nothing more than a collection of information about a particular topic or subject. Database management software permits large amounts of data to be stored efficiently and allows that data to be efficiently sorted, analyzed and reported on. Moreover, many programs permit the user to use the trends and relationships among these data to predict the course of future activity. Although most database programs are somewhat more complicated to use than word processing programs and require some planning and training to use properly, the dividends they pay will become clear as soon as the system is put to use. Most off-the-shelf database software runs in the $200 to $500 price range. Custom designed software for fire investigation and fire reporting costs between $1,000 and $5,000.

An example of a database is a fire department's collection of information about fires and emergency responses obtained from fire incident reports completed by company or chief officers. Most fire investigators are probably familiar with database programs for entering and reporting on such fire incident data. Commercial products for this application abound. An excellent low-cost program is the National Fire Incident Reporting System (NFIRS) software, which is available for the cost of materials and shipping. Copies of the NFIRS software can be obtained by writing the United States Fire Administration, Office of Fire Data and Analysis, 16825 South Seton Avenue, Emmitsburg, Maryland 21727.

The United States Fire Administration also supports a fire investigation software package known as the Arson Information Management System (AIMS). AIMS is a personal computer database system for managing arson investigation data. AIMS is in use by hundreds of fire investigation units nationwide. The profiles of arson-prone buildings are developed by analyzing information about incendiary fires entered in the case file database. This information can then be used to assess the probability that similar buildings entered in another database will be arson targets. Copies of the software are available from the Unites State Fire Administration, Office of Fire Prevention and Control, 16825 South Seton Avenue, Emmitsburg, Maryland 21727.

Spreadsheets

Spreadsheets are programs for collecting and analyzing data. Spreadsheets organize the data into tables. These tables consist of rows and columns. Cells are the junctions between vertical columns and horizontal rows. The data in a cell represents something that column and row have in common. The rows or columns may represent records, classes, or categories or any other piece of information common to other things appearing in the same dimension (horizontal or vertical). The other dimension is used to represent another element, class, category, or record common to the same piece of information.

The value of a spreadsheet is the ability to rapidly calculate and manipulate information which is related logically or mathematically. To do this, however, the user must have some idea what relationships logically exist between the information in the rows and columns. A spreadsheet simply provides a convenient and rational way to organize and

Figure 5.2

Sample Fire Incident Data Spreadsheet

	A	B	C	D	E	F
1	Cause	Year				
2		1991	1990	1989	1988	1987
3	Accidental	3,875	3,602	3,567	3,442	3,712
4	Incendiary	978	898	902	915	765
5	Suspicious	124	136	117	102	181
6	Undetermined	243	256	179	215	175
7	Total	5,220	4,892	4,765	4,654	4,833

manipulate that information. Manipulating data usually means performing logical, statistical, or arithmetic functions on related cells, rows, or columns. Most spreadsheets incorporate features for illustrating these relationship graphically on charts, graphs, and diagrams.

Graphics

Graphics packages include programs for producing, capturing, and manipulating images. These images may be line drawings, three dimensional projections, still photographs, animation or video sequences. Graphics is perhaps the most rapidly advancing area of all computer applications. The growth of this segment is directly related to the power of images to illustrate abstract and complex ideas. Most fire investigators would probably agree that illustrating the relationship between a fire, a building and its contents, and people, including the occupants, witnesses, and firefighters is a complex task. This makes fire investigation an ideal application for graphics programs. The most useful programs for fire investigators are drafting or drawing programs and business graphics.

Computer-aided drafting and design (CADD) packages are excellent programs for producing scaled line drawings of a fire scene. Generally, these programs permit users to produce very sophisticated drawings with great precision. However, many of these programs are highly specialized and require training and practice to use effectively. Simple CADD packages begin at $200. More complex programs for producing three dimension drawings, animated models, and full-color renderings are at the upper end of the cost range at about $4000.

Business graphics programs take the graphics capabilities imbedded in products like spreadsheet programs several steps further. These programs permit the user to produce presentation quality graphics in the form of transparencies, color slides, graphic insects in word processing or DTP documents, and self-running slide shows on the computer screen. Most of these programs also give the user all the tools needed to produce simple line drawingsbut without the same level of sophistication present in the output of CADD programs. Line drawings produced with these programs are usually quite adequate for most fire investigation reports. Many of the business graphics programs also have the capability to convert graphic images captured or produced by other programs for use in their application. A good business graphics package ranges in cost from $200 to $1000. Most programs require a pointing device like a mouse, trackball, or digitizing tablet.

Computer Communications

The telecommunications revolution has followed close on the heels of the information revolution. In fact the two - information and communication - go hand in hand. Once computers were on the scene, it was only a matter of time before it became necessary to connect them to one another to exchange information electronically. Today, hardly any business or agency which uses computers can escape the need to have its computers talk to someone else's.

"Bulletin boards" are software systems that permit users to post notices electronically. To respond to a notice, another user simply identifies which notice he or she wishes to access, and appends the record with a new message or notice. After a time, these records are cleared to make room for others. Many bulletin boards also offer special areas for users to communicate in real-time directly through their computers. In these applications, information is exchanged between the sender and receiver through the host computer with text being directed immediately to the video display terminals of the participants.

Specialized electronic forums for fire service, law enforcement, and fire safety professionals have appeared on mainstream bulletin board and electronic information exchange services in recent years. The most visible of these are managed by national fire service and law enforcement organizations. A few individuals around the country have also setup bulletin boards. The Building and Fire Research Laboratory (BFRL) at the National Institute of Standards and Technology (NIST) operates a special interest bulletin board at no cost other than a toll call to the suburban Washington, D.C. area. The Fire Research Bulletin Board Service (FRBBS) has been in operation since 1986, and has around 200 periodic 'users.'

FIRE MODELING SOFTWARE

A great deal of time over the past decade has been spent improving techniques for predicting fire behavior in simple compartment fires. As the result of these efforts, several excellent programs for modeling fire behavior in single and multiple compartment fires have become widely available. Many of these programs are zone models. A zone model uses the fundamental rules of conservation of mass and energy to calculate the position and travel of the, hot gas layer in compartment fires. In general, these models require users to specify the size and arrangement of compartments, the location and area of vents (windows, doors, mechanical ventilation, and through-penetrations), and the energy release rates, i.e. heat-release-rate (HRR), for the fire(s) to be modeled. The outputs of each model vary, but most predict the time to certain discrete events like flashover or fatal concentrations of specified fire gas species. Most of the available models are programmed to operate on computers using the DOS operating system.

The most widely recognized source of compartment or enclosure computer fire modeling programs is the United States is the Department of Commerce's National Institute of Standards and Technology, Building and Fire Research Laboratory (NIST/BFRL) in Gaithersburg, Maryland. Some of the models which have particular application to fire investigation include the following:

- HAZARD I - user-friendly tire hazard analysis program

- **DETACT-T2** and **DETACT-QS** - predicts detector activation
- ASET - calculates available safe egress time
- CCFM (version VENTS) - multi-compartment zone fire model
- **TENAB** - tenability model
- **EXITT** -exit time simulation program
- **BREAK1** - for glass breakage predictions
- **FPETOOL** - for fire protection engineering cal culations

All of these models can be downloaded from the NIST/BFRL's Fire Research Bulletin Board Service (FRBBS) by calling (301) 921-6302. FRBBS is menu driven and walks the new user through a sign-on and password selection for future access.

Many of the models listed are used to predict the outcome of a given user-defined fire or some aspect of that fire. These models can also provide a useful way of deconstructing or reconstructing the events which led to a specific fire outcome. In the field of fire investigation, the results of a given fire are already known. The difficult task of figuring out what happened requires determiningwhat was there to begin with and how it got the way it was found. Combining the other tools of fire investigation with the power of fire modeling can be a highly effective means of ensuring that fire cause and fire development scenarios are plausible, meaningful, and well-reasoned. Computer fire models are good tools for helping fire investigators understand, *not* determine, what happened during a given fire.

---

[1] FRBBS can be reached at (301) 921-6302 using most commercial communications Your system should be setup for 8 bits, even parity, no stop bit (8-E-0) If you require assistance, the SYSOP (Scot Deal or Charles Arnold) can be reached at (301) 975-6891

Chapter 6 - INFORMATION RESOURCES

A good basic library of texts, handbooks, reports, journals, and periodicals will help the fire investigator stay well-informed about developments in the field and serve as ready references during complex investigations.

THE FIRE INVESTIGATOR'S REFERENCE LIST

The following publications are widely held to be some of the best basic references about fire and arson investigation. Every fire investigation agency should seriously consider acquiring all of the latest of editions of each volume.

FIRE INVESTIGATOR'S REFERENCE LIST

ASTM Committee E-5 on Fire Standards (1990). *Fire Test Standards*, 3rd ed. Philadelphia, PA: American Society for Testing and Materials.

California District Attorney's Association and Aetna Insurance, *Arson Investigation and Prosecution.*

Cole, L. S. *(1985). Investigation of Motor Vehicle Fires*, 2nd ed. Novato, CA: Lee Books.

Cote, A. E., ed. (1991). *Fire Protection Handbook*, 17th ed. Quincy, MA: National Fire Protection Association.

DeHaan, J. D. (1990). *Kirk's Fire Investigation*, 2nd ed. Englewood Cliffs, NJ: Prentice-Hall.

Dils, J. (1990). *Writing Fire and Non-fire Report Narratives.* Pasadena, CA: Jan Dils.

DiNenno, P. J!, ed. (1988). *SFPE Handbook of Fire Protection Engineering*, Quincy, MA: National Fire Protection Association.

International Association of Arson Investigators (1983). *Selected Articles for Fire Investigators*, 5th ed. Louisville, KY: IAAI.

Kodak (1977). *Fire and Arson Photography*, Pamphlet M-67. Rochester, NY: Eastman Kodak Co.

NFPA 921, Guide for Fire and Explosion Investigations (1992). Quincy, MA: National Fire Protection Association.

Shields, T. J. *(1987). Buildings and Fire. New* York: Halsted Press.

Yereance, R. A. (1987). *Electrical Fire Analysis.* Springfield, IL: Charles C. Thomas.

Zulawski, D. E. and Wicklander, D. (1991). *Practical Aspects of Interview and Interrogation.* New York: Elsevier.

JOURNALS AND PERIODICALS

Keeping current in any complex field is a constant challenge. Fortunately, several excellent magazines, periodicals, and journals are available which run articles of interest to the fire investigator. Many of these publications subject published articles to careful editing or peer review to ensure accuracy and integrity and many do not. The following is a partial listing of relevant fire service and fire-related publications available in the United States:

AMERICAN FIRE JOURNAL
9072 E. Artesia Boulevard, Suite 7
Bellflower, CA 90706

BUILDING OFFICIAL AND CODE ADMINISTRATOR
BOCA International, Inc.
4051 W. Flossmoor Road,
Country Club Hills, IL 60478

BUILDING STANDARDS
International Conference of Building Officials
5360 S. Workman Mill Road
Whittier, CA 90601

CALIFORNIA FIREMAN, THE
California State Firemen's Association, Inc.
3246 Ramos Circle
Sacramento, CA 95827

CALIFORNIA STATE FIRE MARSHAL'S NEWS
Special Services Division
7171 Bowling Drive, Suite 600
Sacramento, CA 95823

CHIEF FIRE EXECUTIVE
PTN Publishing Company
445 Broad Hollow Road, #21
Melville, NY 11747

DISPATCH
1325 Pennsylvania Avenue NW
Washington, DC 20004

FIRE AND ARSON INVESTIGATOR, THE
International Association of Arson Investigators
P.O. Box 91119
Louisville, KY 40291

FIRE CHIEF
Communication Channels, Inc.
307 N. Michigan Avenue
Chicago, IL 60601

FIRE CODE JOURNAL
International Fire Code Institute
5360 S. Workman Mill Road
Whittier, CA 90601

FIRE CONTROL DIGEST
3918 Prosperity Avenue, Suite 3 18
Fairfax, VA 22031

FIRE ENGINEERING
Park 80 West
Plaza Two, 7th Floor
Saddle Brook, NJ 07662

FIREFIGHTER'S NEWS
424 S. Rehobeth Road
P.O. Box 165
Milford, DE 19963

FIREHOUSE
PTN Publishing Company
445 Broad Hollow Road
Melville, NY 11747

FIRE TECHNOLOGY
National Fire Protection Association
1 Batterymarch Park
P.O. Box 9101
Quincy, MA 02269

IAFC ON-SCENE
International Association of Fire Chiefs
4025 Fair Ridge Road
Fairfax, VA 22033

INDUSTRIAL FIRE WORLD
208-C Southwest Parkway East
College Station, TX 77842

INTERNATIONAL ASSOCIATION OF
    FIREFIGHTERS MAGAZINE
1750 New York Avenue NW
Washington, DC 20006

INTERNATIONAL CONNECTIONS
International. Association of Fire Chiefs
4025 Fair Ridge Road
Fairfax, VA 22033

MINNESOTA FIRE CHIEF
3 17 Owasso Boulevard South
St. Paul, MN 55 113

NATION'S CITIES WEEKLY
National League of Cities
1301 Pennsylvania Avenue NW
Washington, DC 20004

NFPA JOURNAL
National Fire Protection Association
1 Batterymarch Park
P.O. Box 9101
Quincy, MA 02269

OREGON FIRE SERVICE GATED WYE
Oregon State Fire Marshal's Office
4760 Portland Road NE
Salem, OR 97305

PUBLIC FIRE EDUCATION DIGEST
Oklahoma State University
2004 S. Deerborn Circle
Columbia, MO 65203

SIZE-UP
New York State Association of Fire Chiefs
1670 Columbia Turnpike
Castleton, NY 12033

SPEAKING OF FIRE
International Fire Service Training Association
Fire Protection Publications
Oklahoma State University
Stillwater, OK 74078

THE VOICE
International Society of Fire Service Instructors
30 Main Street
Ashland, MA 01721

PROCEEDINGS

Much of the important work done at the frontier of fire science does not appear in the better known and more mainstream fire protection publications. Most academic conferences and technical symposiums do, however, publish documents such as proceedings. These publications serve as conduits for the papers presented at these meetings and conferences. Many of these meetings convene at regular intervals, others are far more *adhoc,* occurring only once or periodically but irregularly. The annual National Arson Forum meetings sponsored by the Insurance Committee for Arson Control and the Worcester Polytechnic Institute sponsored Conference on Fire Safety Design in the 21st Century are examples of such symposiums.

ELECTRONIC INFORMATION SERVICES

Chapter 5 contains information about several electronic information services of particular interest to fire investigators. Many of the mainstream information services give users the ability to set up clipping files newspaper and wire service articles about topics of interest by using keywords to search the text of each story as it appears on the network. This can be a particularly efficient way of gaining up-to-date information about emerging trends in the field or significant events around the nation or the world.

Specialized Electronic Reference Sources

Many libraries have begun providing access to specialized electronic reference sources, occasionally charging a small fee to cover their on-line charges. Some of the more sophisticated services give users up-to-date access to legal decisions, law review articles, scientific and technical journals, social science abstracts, medical journals, patent and trademark files, and dissertation abstracts. In fact, most major reference sources in print have begun to appear in electronic versions.

Fire Service Bulletin Boards

In recent years, fire-oriented bulletin board services have begun to emerge. The International Association of Fire Chiefs and the National Volunteer Fire Council operate two such services. The IAFC information service, ICHIEFS, operates as a special interest forum on another on-line service called Connect™ . FireWatch, the NVFC forum, operates as stand-alone system from a computer located in Colorado. As computer technology becomes increasingly ubiquitous one can only expect the number of fire-related electronic information services to expand as well. Contacts are as follows:

ICHIEFS
Ken Westlund
National Volunteer Fire Council
1422 East 110th Place
Northglen, Colorado 80233
(303) 452-9992

FireWatch
Ann Swing
International Association of Fire Chiefs
4025 Fair Ridge Road
Fairfax, Virginia 22033
(703) 273-0911

UNITED STATES FIRE ADMINISTRATION PUBLICATIONS

The following United States Fire Administration publications provide information useful to fire and arson investigators. A catalog containing all United States Fire Administration publications, titled *Resources on Fire* (FA-102), is available by contacting the United States Fire Administration, 16825 South Seton Avenue, Emmitsburg, Maryland **21727, (301) 447-1000.**

Fire and Arson Investigator's Field Index Directory (FA-91) - This publication serves as a compendium of expert contacts in federal agencies, government and private research laboratories, codes and standards organizations, trade associations, product manufacturers and distributors, insurance companies and organizations, law enforcement agencies and organizations, and other valuable contacts for technical information and resources for fire and arson investigators handling complex cases.

Arson Resource Directory (FA-74) - This directory provides a comprehensive listing and explanation of the roles of the various organizations and agencies across the nation involved in the fight to control arson. It includes names and contact information for county and metropolitan fire and arson investigation units, state fire marshals offices, state training organizations, National Fire Incident Reporting system coordinators, federal agencies, and other organizations such as the Federal Arson Task Force and International Association of Arson Investigators, and other fire service and insurance organizations.

Arson Control Directory - This directory is produced annually through a cooperative agreement between the United States Fire Administration, the Insurance Committee for Arson Control, National Association of State Fire Marshals, and the International Association of Arson Investigators working through the National Arson Forum. It details the people, groups, and organizations working to reduce the incidence of arson in the United States.

Public Fire Education Today: Fire Service Programs Across America (FA-98) - This publication identifies and describes successful public fire education programs in all 50 states and the District of Columbia. Several good examples of anti-arson and juvenile firesetter programs are included in the cases presented.

Rural Arson Control (FA-87) - This study report outlines the difficulties confronting rural communities in the fight against arson and recommends solutions to help these communities win the war against incendiarism.

Establishing an Arson Strike Force (FA-88) - This report describes the Arson Strike Force concept for investigating fires and prosecuting the crime of arson. This model has proven to be one of the most successful programs for combatting arson. The report outlines how to establish and administer and arson strike force in your community.

Arson Prosecution: Issues and strategies (FA-78) - This report describes the recommended methods for successfully developing and prosecuting arson cases. It details the roles of prosecutors, eyewitnesses, expert witnesses, investigators, and different types of evidence and emphasizes how to present each element of the prosecution to build a credible and convincing case.

Preadolescent Firesetter Handbook (3 volumes) (FA-80, FA-82 and FA-83) - These manuals present an overview of juvenile firesetter behavior and outline successful counseling and intervention programs and strategies for three different age groups, 0-7, 7-13, and 14-18. Sample interview and evaluation forms are provided to illustrate key concepts. And contacts are provided for existing programs in communities nationwide.

Special Report - Arson in America: A Profile of 1989 NFIRS - This report examines the scope of the incendiary fire problem in America through an analysis of National Fire incident Reporting System (NFIRS) data for calendar year 1989. NFIRS data is based on reports collected from over 10,000 fire departments in 40 states, and 27 metropolitan areas which cooperate in the National Fire Incident Reporting System.

A View of Management in Fire Investigation Units (FA-93) - Particular management practices from several local fire investigation units are highlighted. Factors affecting the current state of fire investigation are described, including the trend away from joint police and fire investigation teams, the impact of drug wars on the quantity and danger of investigations, the need for management training for investigation managers, and inadequate data collection and application.

In addition to these publications, the United States Fire Administration maintains an Arson Resource Center in the National Emergency Training Center Learning Resources Center. Expert technical reference librarians are on-call to assist local fire investigations in finding information and specific fire and arson investigation topics. The Arson Resource Center can be reached by calling toll-free (800) 638-1821.

BIBLIOGRAPHY

Chapter 2 - BASIC TOOLS AND EQUIPMENT

Ayers, J. H., and associates (May 1983). The Fire Investigation Van and Equipment, National Fire Academy Fire/Arson Investigation Research Paper. Emmitsburg, MD: United States Fire Administration.

Wittig, C. (March 1991). "Hazards for the Fire Investigator," *Fire and Arson Investigator,* Vol. 4 1, No. 3, pp. 22-23.

Zilke, F. and Wachowiak, M. "The Value of Photography in a Small Fire Department," *National Fire and Arson Report,* Vol. 8, No. 5, p. 6, 11.

Chapter 3 - ACCELERANT DETECTION

Berluti, A. F. (Dec. 1990). "Arson Investigation: Connecticut's Canines," *Police Chief,* pp. 39-45.

Berluti, A. F. (June 1989). "Sniffing Through the Ashes," *Fire and Arson Investigator,* Vol 39, No. 4 (pp. 31-35).

Byron, M. M. (1982). "Three Commonly Used Accelerant Detectors," *National Fire and Arson Report,* Vol. 1, *No. 2,* pp. 6, 14.

Campbell, D. H. "Is This Sniffer Here to Stay?", *National Fire and Arson Report,* Vol. 8, No. 5, p. 7.

Connecticut State Police (Aug. 1988). Canine Accelerant Detection Program. Meriden, CT: Connecticut State Police.

Custer, R. L. P. (1991). "Fire Loss Investigation," in *Fire Protection Handbook,* 17th ed., Cote, A. E., ed. Quincy, MA: National Fire Protection Association.

Dorriety, J. K. (June 1991). "Accelerant Detector Dogs are Valuable Investigation Tools," *Fire Chief,* Vol. 35, *No. 6,* p. 50.

Meyer, R. E. (Oct.-Dec. 1981). "Evaluation of Three Commonly Used Hydrocarbon Detectors," *Fire and Arson Investigator,* Vol. 32, No. 2, pp. 35-38.

Pragmatics, Inc. (n.d.). Operation and Applications of the Model 653 MkII "Serviceman Surveyor" and Models 920/620 "Fire Investigators"

Telephone interview with Richard Motsinger, Pragmatics, Inc., August 19, 1992. Telephone interviews with Michael Higgins, Laboratory Instrument Services, August 19, 1992 and September 1, 1992.

Telephone interviews with Sergeant James R. Butterworth, Connecticut State Police, July 1989.

Chapter 4 - COLLECTION, PACKAGING, AND ANALYSIS OF EVIDENCE

Alford, C. and associates (May 1983). Understanding the Gas Chromatograph, National Fire Academy Fire/Arson Investigation Research Paper. Emmitsburg, MD: United States Fire Administration.

Bertsch, W., Zhang, Q. W., and Holzer, G. (Sept. 1990). "Using the Tools of Chromatography, Mass Spectrometry, and Automated Data Processing in the Detection or Arson," *Journal of High Resolution Chromatography,* Vol. 13, pp. 597-604.

Byron, M. M. (Dec. 1990). "Adequate Sampling of Fire Scenes," *Fire and Arson Investigator,* Vol. 41, No. 2, pp. *48-49.*

Byron, M. M. (Sept. 1989). "Inadequate Review of Field Data," *Fire and Arson Investigator,* Vol. 40, No. 1, pp. 61-62.

Chasteen, C. E. (1992). Guide to the Collection, Packaging, Submission, and Analysis of Evidence from Suspicious Fires. Tallahassee, FL: Division of State Fire Marshal, Bureau of Fire and Arson Laboratory.

"Cold Hands, Bright Snow, Dead Batteries: Challenges of Cold-Weather Photography," *Fire and Arson Investigator,* Vol. 40, No. 2, (Sept. 1989), p. 12.

Dietz, W. R. (June 1991). "Physical Evidence of Arson: Its Recognition, Collection and Packaging," *Fire and Arson Investigator,* Vol,. 41, No. 4, pp. 33-39.

Eaton, T. E. (March 1990). "Fire Transfer," *Fire and Arson Investigator,* Vol. 40, No. 3, pp. 42-48.

Galvin, M. and Toscano, J. P. (Dec. 1990). "The New Fire Triangle: Putting the Prosecutor on the Team," *Police Chief,* pp. 50-52.

Henderson, R. W. (1989). "A Simple Introduction to Gas *Chromatography, "National Fire and Arson Report,* Vol. 7, No. 3, pp. 3, 14.

International Association of Arson Investigators Forensic Science Committee (June 1991). "Standard Test Method for: Flammable or Combustible Liquid Residues in Extracts from Samples of Fire Debris by Gas Chromatography," *Fire and Arson Investigator,* Vol. 41, No. 4, pp. 51-55.

international Association of Arson Investigators Forensic Science Committee (Dec. 1990). "Forensic Science Committee Position on Comparison Samples," *Fire and Arson Investigator,* Vol. 41, No. 2, pp. 50-51.

Lentini, J. J. (March 1991). "ASTM Adopts Fire Debris Analysis Standards," *Fire and Arson Investigator,* Vol. 41, No. 3, p. 16.

Lentini, J. J. (June 1989). "Control Samples, A Reply," *Fire and Arson Investigator,* Vol. 39, No. 4, pp. 39-40.

Meal, L. (Sept. 1989). "An Alternative to Gas Chromatography in the Analysis of Fire Debris," *Fire and Arson Investigator,* Vol. 39, No. 1, pp. 55-56.

Nowicki, J. (Sept. 1990). "An accelerant Classification Scheme Based on Analysis by Gas Chromatography/ Mass Spectrometry (GC-MS)," *Journal of Forensic Sciences,* Vol 35, No. 5, pp. 1064-1086.

O'Donnell, J. F. (Sept. 1990). "Accelerant Residue Analysis by a Combined Solvent Extraction-Heated Headspace Technique," *Fire and Arson Investigator,* Vol. 41, No. 1, pp. 4-5.

O'Donnell, J. F. (June 1989), "Interferences from Backgrounds in Accelerant Residue Analysis," *Fire and Arson Investigator,* Vol. 39, No. 4, pp. 25-27.

Posey, E. P. (June 1990). "Attention IAAI Members, Do You Know About ASTM's E-30 Committee on Forensic Sciences?", *Fire and Arson Investigator,* Vol. 40, No. 4, pp. 52-53.

Sanders, D. (March 1991). "The Investigation of Fires Through Photographs," *Fire and Arson Investigator,* Vol. 41, No. 3, p. 60.

Sanderson, J. L., Balliet C. A, and Balliet, M. A. (March 1990). "Sampling Techniques for Accelerant Residue Analysis," *Fire and Arson Investigator,* Vol. 40, No. 3, pp. 35-40.

Thompson, V. A. (Sept. 1989). "Slides, Color Prints or Color Negatives," *Fire and Arson Investigator,* Vol. 40, No. 1, p. 48.

Tsaroom, S, Elkayam, R., and Natayan, S. (March 1990). "Evaporation of Gasoline and Kerosene from Polyethylene Containers," *Fire and Arson Investigator,* Vol. 40, No. 3, pp. 21-23.

Wagner, R. W. (March 1991). "The Investigation of Fire Through Photographs! Yes, Impossible!", *Fire and Arson Investigator,* Vol. 41, No. 3, pp. 58-59.

Waters, L. V. (March 1991). "Adequate Analysis of Laboratory Data," *Fire and Arson Investigator,* Vol. 41, No. 3, pp. 61-62.

Chapter 5 - COMPUTER SYSTEMS

Beering, P. S. (Feb. 1992). "High-Tech Hunting and Gathering: Using Computers in Fire Investigations," *Firehouse,* pp. 52-54.

Beyler, C. R. (1991). "Introduction to Fire Modeling," in *Fire Protection Handbook,* 17th ed., Cote, A. E., ed. Quincy, MA: National Fire Protection Association.

Bukowski, R. W. (March 1992). "Analysisof the Happyland Social Club Fire with HAZARD *I,"* *Fire and Arson Investigator,* Vol. 42., No. 3, pp. 36-47.

Clarke, F. B. (1991). "Fire Hazard Assessment," in *Fire Protection Handbook,* 17th ed., Cote, A. E., ed. Quincy, MA: National Fire Protection Association.

Daugenti, J., and associates (May 1982). The Evaluation of Computers in Arson Investigation, National Fire Academy Fire/Arson Research Paper. Emmitsburg, MD: United States Fire Administration.

DiNenno, P. (1991). "The Future of Fire Modeling,"in *Fire Protection Handbook,* 17th ed., Cote, A. E., ed. Quincy, MA: National Fire Protection Association.

Fire Research Bulletin Board Service (FRBBS), electronic information service. Gaithersburg, MD: National Institute of Standards and Technology, Building and Fire Research Laboratory.

Hall, J. R., Jr. (1991). "Use of Fire Loss Information," in *Fire Protection Handbook,* 17th ed., Cote, A. E., ed. Quincy, MA: National Fire Protection Association.

Nelson, H. E. (1991). "Application of Fire Growth Models to Fire Protection Problems," in *Fire Protection Handbook,* 17th ed., Cote, A. E., ed. Quincy, MA: National Fire Protection Association'.

Nelson, H. E. (July 1989). "Science in Action: An Engineering View of the Fire at the First Interstate Bank Building," *Fire Journal,* Vol. 83, No. 4, pp. 28-34.

"New Computer Application for Investigators," *National Fire and Arson Report,* Vol. 8, No. 2, p. 6.

Perroni, C. A. (July 1989). "Investigating Fires the Scientific Way," *Fire Journal,* Vol. 83, No. 4, pp. 24-27.

Peterson, C. E. (1991). "Collecting Fire Data," in *Fire Protection Handbook,* 17th ed., Cote; A. E., ed. Quincy, MA: National Fire Protection Association.

Walton, W. D. and Budnick, E. K. (I 991). "Deterministic Fire Models," in *Fire Protection Handbook,* 17th ed., Cote, A. E., ed. Quincy, MA: National Fire Protection Association.

Watts, J. M., Jr. (1991). "Microcomputer Applications in Fire Protection," in *Fire Protection Handbook,* 17th ed., Cote, A. E., ed. Quincy, MA: National Fire Protection Association.

APPENDIX A
FIRE INVESTIGATION TOOLS AND EQUIPMENT LISTS

HAND TOOLS

Chisels
  Cold chisel(s)
  Wood chisel(s)
Crowbar/prybar
Cutter(s), cutting tools
  Bolt cutters
  Carpet cutters
  Pipe cutter
  Pocket knife
  Razor blades
  Tin snips
  Utility knife
Drill with wood and metal bits
Electrician's tools
  Circuit continuity tester
  Electrical tape
  Fuse puller
  Ground fault/phase tester
  Volt/Ohm/Amp meter
  Wire cutters
  Wire gauge Wire stripper
Hammers
  Ballpeen
  Claw
Hand axe/hatchet
Hand trowel
Level (torpedo and bar)
Magnet
Magnetic compass
Measuring tools
  Folding rule
  Measuring wheel
  Ruler (12-in, 14-in, 18-in)
  Tape (25-ft, 50-ft, 100-ft)
Paint brush(es)
Pliers
  Lineman's
  Needle-nose
  Sidecutting
  Slip-lock
  Vise-Grip™
Plumb bob
Putty knife/scraper(s)
Saws
  Crosscut
  Hack
  Keyhole
  Scroll

Screwdriver(s)
  Bit driver with assorted bits
  Phillips head
  Straight blade
Screws, nails, tacks, staples
Wrenches
  Adjustable
  Open-end or box
  Pipe

EVIDENCE TOOLS AND EQUIPMENT

Brush(es)
Containers, packaging materials
  Cardboard boxes, assorted sizes
  Document protector(s)
  Envelopes, kraft paper or tyvek™, assorted sizes
  Evidence tape (tamper indicating)
  Glass jars w/ screw-top lids, assorted sizes
  Glass vials w/ screw-top lids
  Kraft paper
  Masking tape
  Paper bags
  Plastic (polyester, nylon, or Kapak™) bags, heavy duty (+3-mil)
  Steel paint cans w/ friction lids, assorted sizes
  String
  Transparent or cellophane tape
  Vermiculite
Drawing equipment
  Compass (drafting)
  Drafting templates, architectural
  Pens, pencils
  Protractor
  Scale, architectural
  Scale, engineering
  Straight edge
  Sketch pads, quadrille ruled
Fingerprint kit
Latex rubber gloves
Magnifying glass
Marking equipment
  Evidence tags/forms
  Marking pens, indelible ink
  Scribe, etching tool
Measuring equipment
  Magnetic compass
  Scale or balance
  Range finder
Meters

Barometer
Combustible gas/vapor detector
Thermometer, dry bulb
Thermometer, wet bulb
Velometer
pH
Notebooks
Recording equipment, sound
   Blank cassette tapes
   Tape recorder
Sample collection equipment
   Activated charcoal
   Blotter strips
   Cotton-tipped swabs
   Cotton baby diapers, clean, unused
   Dental tools
   Disposable sterile pipets with suction bulbs
   Dust pans
   Eye dropper
   Forceps
   Scalpel
   Sifting screens or sieve
   Syringes
   Tongs
   Tweezers
UV (black) light

PHOTOGRAPHIC AND VIDEO EQUIPMENT

Batteries, spare
   Camera
   Flash
Camera
   35 mm automatic
   35-mm SLR
   Instant
Film, spare
Flash attachment
Instant camera macro attachment
Lenses
   70-200 mm zoom
   28-70 mm zoom
   50 mm
Meter
   Flash
   LightMotor winder
Photographic background
Photographic flags/markers
Photographic logs
Photographic ruler
Protective carrying cases
Tripod
Videotape boom microphone
Videotape lighting attachment
Videocassettes, blank, spare

PERSONAL PROTECTIVE EQUIPMENT

Ballistic vest
Brush firefighting coat
Brush firefighting pants
Boots, steel toe/shank safety
Chemical splash protective coveralls
Coveralls
Dust mask(s)
Filter mask respirator
Firefighter's boots
Firefighter's gloves
Firefighter's turnout coat
Firefighter's turnout pants
Goggles/safety glasses
Hardhat
Law enforcement equipment
   Handcuffs/case
   Handgun, holster, ammunition
   Gun cleaning kit
Nomex^{TM}/PBI hood
Respirator filters, organic vapor/acid gas/HEPA
Self-contained breathing apparatus
Spare SCBA bottle(s)
Structural firefighter's helmet
Suspenders
Work gloves

COMMUNICATION EQUIPMENT

Cellular telephone
Facsimile machine
   Cellular
   Office
Portable radio
Pager
   Radio
   Telephone
Portable computer w/ modem
Public address system

VEHICLE EQUIPMENT

AC/DC charger
First aid kit
Flood lights
Fusees/light sticks
Jumper cables
Mobile Radio
Portable fire extinguisher
Public address system
Reflective markers
Siren

Spot light(s)
Warning lights

MISCELLANEOUS TOOLS AND EQUIPMENT

Barricade tape
Broom(s)
Buckets
Claw tool
Cleaning rags
Entrenching tool
Extension cords
Flashlights
Hand cleaner
Hand lanterns
Hydrant wrench
Ladder, extension
Ladder, step
Light stands
Pick
Portable generator
Portable lights
Rope
Shovel, chisel point
Shovel, scoop
Shovel, square head
Spade
Squeegee
Tarps/salvage covers

APPENDIX B
GLOSSARY OF FORENSIC SCIENCE TERMS

Source: International Association of Arson Investigators Forensic Science Committee (Dec. 1989), "Glossary of Terms Related to Chemical and Instrumental Analysis of Fire Debris," *Fire and Arson Investigator,* Vol. 40, No. 2, (pp. 25-34).

-A-

Absorption: 1) A mechanical phenomenon wherein one substance penetrates into the inner structure of another, as in absorbent cotton or a sponge. 2) An optical phenomenon wherein atoms or molecules block or attenuate the transmission of a beam of electromagnetic radiation.

Accelerant: Any material used to initiate or promote the spread of a fire. The most common accelerants are flammable or combustible liquids. Whether a substance is an accelerant depends on its chemical structure, but on its use.

Acetone: The simplest ketone. A highly flammable, water soluble solvent. Flash point of 0° F. Explosive limits of 2.6% to 12.8%.

Adsorption: The adherence of atoms, ions, or molecules of a gas or liquid to the surface of another substance. Finely divided or microporous materials having a large active surface area, activated carbon, activated alumina and silica gel.

Alcohol: An organic compound having a hydroxyl (-OH) group attached. The lower molecular weight alcohols, methanol (CH_3OH), ethanol (C_2H_5OH), and propanol (C_3H_7OH), are water soluble.

Aliphatic: One of the main groups of hydrocarbons characterized by the straight or branched chain arrangement of constituent atoms. Aliphatic hydrocarbons belong to three subgroups: (1) alkanes or paraffins, all of which are saturated and comparatively unreactive, (2) the alkenes or alkadienes which are unsaturated (containing double [C=C] bonds) and more reactive, and (3) alkynes, such as acetylene (which contain a triple [C=C] bond).

Alkane: An aliphatic hydrocarbon having the chemical formula CnH_2n+2. A normal alkane, or n-alkane is one which does not have a branched carbon backbone. An isoalkane has a branched, rather than a straight chain, carbon backbone. Alkanes are also known as paraffins. The simplest alkanes are named as follows:

CH_4 - methane	C_8H_{18} - octane
C_6H_{14} - hexane	C_4H_{10} - butane
C_2H_6 - ethane	C_9H_{20} - nonane
C_7H_{16} - heptane	C_5H_{12} - pentane
C_3H_8 - propane	$C_{10}H_{22}$ - decane

Alkene: A straight chain, unsaturated compound of the olefin series which has the generic formula CnH_2n, having at least one double [C=C] bond. (See Aliphatic).

Alkyl Group: A functional group having the formula CnH2n+1 which may be attached to certain elements such as lead, silicon, or to other organic chemicals.

Alkyne: An unsaturated aliphatic hydrocarbon characterized by the presence of a triple [C=C] bond. The generic formula for an alkyne is CnH2n-2. The most important member of this group is acetylene, HCCH, the first member of the series.

Alloy: A solid or liquid mixture of two or more metals, or of one or more metals with certain nonmetallic elements, as in brass, bronze, or carbon steel.

Ambient: Pre-existing or normal environment.

Aromatic: An organic compound having as part of its structure a benzene ring. (See Benzene). The term "aromatic" as used in the fragrance industry is used to describe essential oils, which are not necessarily aromatic in the chemical sense.

Arson: The crime of intentionally setting fire to a building or other property. This is a legal definition which may vary depending on the laws of a specific state.

Atom: The smallest unit of an element which still retains the chemical characteristics of that element. An atom is made up of protons and neutrons in a nucleus surrounded by electrons. A molecule of water (H_2O) consists of two atoms of hydrogen and one atom of oxygen.

Atomic Absorption: An analytical technique, used to determine the elemental composition and the concentration of many metals and other inorganic elements. The material being analyzed, generally in solution, is atomized, or broken up into individual atoms, usually by the action of extreme heat in a flame or small furnace. The ability of the atomized material to absorb characteristic wavelengths of visible or ultraviolet light is then measured using a spectrophotometer.

Atomize: 1) To break into discrete atoms, usually by the application of extreme heat, as in atomic absorption. 2) To break a liquid into tiny droplets, as occurs in fuel injected engines or in the production of aerosol sprays.

Attenuation: An adjustment of the signal amplifier response which results in the reduction of the electronic signal.

Azeotrope: A mixture of two or more compounds which has a constant boiling point. The composition of the vapor above the azeotropic mixture has the same relative concentrations of compounds as does the boiling liquid. Azeotropic mixtures cannot be separated by fractional distillation.

-B-

Benzene: A hexagonal organic molecule having a carbon atom at each point of the hexagon, and a hydrogen atom attached to each carbon atom. Molecules which contain a benzene ring are known as aromatic. Benzene boils at 80° C, and has a flash point of 12° F(-11° C). The explosive limits are 1.5% to 8% by volume in air.

BTU: British Thermal Unit. The amount of heat energy required to raise the temperature of one pound of water one degree Fahrenheit. This the accepted standard for the comparison of heating values of different fuels. One BTU equals 252 calories.

Burning: Normal combustion in which the oxidant is molecular oxygen.

Burning Rate: The rate at which combustion proceeds across a fuel. A specialized use of this term describes the rate at which the surface of a pool of burning liquid recedes. For gasoline, this rate is reported to be approximately 1/4 inch per minute.

Butane: A fuel gas having the formula C_2H_{10}. A constituent of LP gas. One pound of liquid butane produces 6.4 cubic feet of gas. One gallon of liquid butane weighs 4.87 pounds and produces 31 cubic feet of gas. One cubic foot of butane gas produces 3266 BTU's.

-C-

Calorie: The amount of energy required to raise the temperature of one gram of water one degree Centigrade. One calorie equals 0.004 BTU's One BTU equals 252 calories.

Capillary: A narrow boreglass tube. Capillary column gas chromatography employs glass tubes having an inside diameter of approximately .2 to .5 millimeters and a length of 3 to 300 meters. The walls of a capillary column are coated with an adsorbent or adsorbents medium (a liquid phase in which the sample dissolves).

Carbon: The element upon which all organic molecules are based. Carbon has an atomic weight of 12.00, and occurs elementally in these forms: diamond, graphite, and amorphous carbon such as coal or carbon black.

Carbon Dioxide: A molecule consisting of one atom of carbon and two atoms of oxygen which is a major combustion product of the burning of organic materials. Carbon dioxide (CO_2) is the result of complete combustion of carbon. In the gaseous form, CO_2 is used as a fire extinguisher. In the solid form, CO_2 is known as dry ice. CO_2 is heavier than air, with a vapor density of 1.53 (air=1.00).

Carbon Disulfide: A highly flammable nonpetroleum solvent used extensively for gas chromatography because of its relatively low signal generated in a flame ionization detector. Carbon disulfide has the formula CS_2. Reagent grade CS_2 has an odor similar to rotten broccoli, and can be ignited by contact with boiling water. It burns with a blue flame, producing CO_2 and SO_2 (sulfur dioxide). The explosive limits of CS_2 are 1 to 50%. CS_2 has a flash point of -22° F.

Carbon Monoxide: A gaseous molecule having the formula CO, which is the product of incomplete combustion of organic materials. Carbon monoxide has an affinity for hemoglobin approximately 200 times stronger than oxygen's and is highly poisonous. CO is a flammable gas which burns with a blue flame, and has explosive limits of 12% to 75%. Carbon monoxide has approximately the same vapor density as air, 0.97 (air=1.00).

Carbon Tetrachloride: A nonflammable liquid having the formula $CC_{,,,}$ formerly used as a fire extinguisher, and still used as a solvent and cleaning agent. Carbon Tetrachloride boils at 77° C.

Chain Reaction: A self-propagating chemical reaction in which activation of one molecule leads successfully to activation of many others. Most, perhaps all, combustion reactions are of this kind.

Chemical Change: Rearrangement of the atoms, ions or radicals of one or more substances, resulting in the formation of new substances, often having entirely different properties.. Also known as a chemical reaction.

Chemistry: A basic science concerned with (1) the structure and behavior of atoms (elements); (2) the composition and properties of compounds; (3) the reactions that occur between substances, and the resultant energy exchange and (4) the laws that unite these phenomena into a comprehensive system.

Chromatography: A chemical separation procedure which separates compounds according to their affinity for an adsorbent or absorbent material. Chromatography includes Thin Layer Chromatography (TLC), Liquid Chromatography (LC), Gas Chromatography (GC) (sometimes called Gas Liquid Chromatography or GLC) and High Performance Liquid Chromatography or High Pressure Liquid Chromatography (HPLC).

Comparison Sample: 1) A sample of material collected from a fire scene which is, to the best of the investigators knowledge, identical in every respect to a sample suspected of containing accelerant, but which does not contain accelerant. 2) A sample of suspected accelerant submitted to the purpose of comparing with any accelerant separated from a debris sample. (See Control Sample).

Combustible Liquid: A liquid which is capable of forming a flammable vapor/air mixture. All flammable liquids are combustible. Whether a liquid is combustible or flammable depends on its flash point and on the agency definition relied upon. The Coast Guard classifies all liquids with a flash point below 80° F as flammable. The NFPA uses 100° F.

Combustion: An exothermic chain reaction between oxidizing and reducing agents, or between oxygen and fuel. Combustion may occur with any organic compound, or with certain combustible elements such as hydrogen, sulfur, and finely divided metals.

Component: 1) One of the elements or compounds present in a system such as a phase, a mixture, a solution, or a suspension in which it may or may not be uniformly dispersed. 2) A compound or a group of unresolved compounds represented by a peak on a chromatogram.

Compound: A chemical combination of two or more elements, or two or more different atoms arranged in the same proportions and in the same structure throughout the substance. A compound is different from a mixture in that the components of a mixture are not chemically bonded together. For example, a flash may contain two volumes of hydrogen (H_2) gas and one volume of oxygen (O_2) gas. A different flash might contain only water vapor (H_2O). In the first case, two gases are mixed. In the second case only one gas is present.

Concentration: The amount of substance in a stated unit of a mixture or solution. Common methods of stating concentration are per cent by weight, per cent by volume, or weight per unit volume. (e.g. parts per million, billion, etc.)

Conduction: Passage of heat from one material to another by direct contact.

Conductivity: The ability of a material to transfer energy from one place to another. Thermal conductivity describes a substance's ability to transmit heat. Electrical conductivity describes a substance's ability to transmit electrical current. Conductivity is the opposite of resistivity.

Control Sample: A sample of material which is known to be identical to a sample suspected of containing accelerant in every regard, except that the control sample does not contain accelerant. A known blank sample. In practical terms, control samples do not exist in the setting of a fire

scene because 1) exact conditions during the fire cannot be duplicated from location to location, and 2) exact conditions of the substrate, i.e., carpet, etc., are not known. Carpet in Room A may have had various items spilled on it during use. Carpet B (ten feet away) may be brand new. (See Comparison Sample).

Convection: Transfer of heat by the movement of molecules in a gas or liquid, with the less defense fluid rising. The majority of heat transfer in a fire is by convection.

Corrosion: The degradation of metals or alloys due to reaction with their environment. It is accelerated by acids, bases or heat.

Cracking: A refining process involving decomposition and molecular recombination of organic compounds, especially hydrocarbons obtained by distillation of petroleum, by means of heat, to form molecules suitable for various used such as motor fuels, solvents or plastics. Cracking takes place in the absence of oxygen,.

-D-

Deflagration: Vigorous burning with subsonic flame propagation. (See Detonation)

Desorption: The process of removing an adsorbed material from the solid on which it is adsorbed. (See Adsorption)

Detonation: An exothermic chemical reaction which propagates through reactive material at supersonic speed.

Diesel Fuel: Diesel Fuel consists mostly of hydrocarbons ranging from C_{10} through C_{24}. The composition of diesel fuel may vary with changes in latitude or changes in season. This variability is provided by the refinery to control the volatility of the product. In order to be identified as diesel fuel, a sample extract must exhibit a homologous series of five or more consecutive normal alkanes ranging from C_{12} through C_{22}. Diesel fuel has a flash point of 120 to 160° F and explosive limits of 0.7% to 5%. Many states specify a minimum flash point for diesel fuel.

Distillation: A separation process in which a liquid is converted to a vapor, and the vapor is then condensed back to a liquid. The usual purpose of distillation is separation of the compounds of a mixture. Steam distillation separates all water insoluble liquids from solids and water soluble compounds in a mixture.

Drying Oil: An organic liquid which, when applied as a thin film, readily absorbs oxygen from the air and polymerizes to form a tough elastic film. Linseed, tung, soybean and castor oils are drying oils. 'Under certain conditions, usually involving large surface areas and insulation, such as a pile of rags soaked with drying oils, spontaneous heating may occur.

-E-

Electron: A negatively charged subatomic particle which circles the nucleus of the atom in a cloud. Most chemical reactions involve the making and breaking of bonds held together by the sharing of electrons.

Electron Capture Detector (ECD): A type of gas chromatographic detector which is sensitive to halogenated hydrocarbons and other molecules capable of easily gaining an electron. Electron capture is not generally used for hydrocarbon detection.

Element: One of 106 presently known kinds of substances that comprise all matter at and above the atomic level.

Elution: The process of removing adsorbed materials from the surface of an adsorbent such as activated charcoal. The solvent in this process is called the eluant.

Emission Spectroscopy: The study of the composition of substances and identification of elements by observation of the wave lengths of radiation emitted by the substance as it returns to a normal state after excitation by an external source. Generally used for elemental analysis.

Emulsion: A stable mixture of two or more immiscible liquids in suspension.

Endothermic Reaction: A chemical reaction which absorbs heat.

Ethane: A simple alkane having the formula C_2H_6. A minor component of natural gas. Its explosive range is 3% to 12.5%. Ethane has approximately the same vapor density as air.

Ethanol: Ethyl alcohol. Grain alcohol. Flammable, water soluble alcohol. Flash point of 55° F. Explosive limits of 3.3% to 19%.

Ether: Diethyl ether, ethyl ether. A highly flammable solvent which can form explosive peroxides when exposed to air. Flash point of -49F. Explosive range of 1.85% to 48%.

Ethylbenzene: A component of gasoline, but also a major breakdown product or pyrolysis product given off when certain polymers are heated.

Eutectic: The lowest melting point of an alloy or solution of two or more substances (usually metals) that is obtainable by varying the percentage of the components. Eutectic melting sometimes occurs when molten aluminum or molten zinc comes in contact with solid steel or copper.

Evaporation: Conversion of a liquid to the vapor state. Set also Vaporization.

Evaporation Rate: A measure of the quantity of a liquid converted to vapor in a unit of time. Among single component liquids, the rate varies directly with the surface area, the temperature and the vapor pressure, and inversely with the latent heat of vaporization of the liquid.

Exothermic Reaction: A chemical reaction which evolves heat. Combustion reactions are exothermic.

Explosion: The sudden conversion of chemical energy into kinetic energy with the release of heat, light and mechanical shock.

Explosive Limit: Flammability limit. The highest or lowest concentration of a flammable gas or vapor in air that will explode or burn readily when ignited. This limit is usually expressed as a volume percent of gas or vapor in air.

Explosive Range: Flammability range. The set of all concentrations between the upper and lower, explosive limits of a particular gas or vapor.

Extraction: A chemical procedure for removing one type of material from another. Extraction is generally carried out by immersing a solid in a liquid, or by shaking two immiscible liquids together, resulting in the transfer of a dissolved substance from one liquid to the other. Solvent extraction is one of the primary methods of sample preparation in arson debris analysis.

-F-

Fire: The light and heat manifested by the rapid oxidation of combustible materials. A flame may be manifested but is not required.

Fire Point: The temperature, generally a few degrees above the flash point, at which burning is self-sustaining after removal of an ignition source.

Fire Tetrahedron: Fuel, heat, oxygen and a chemical chain reaction.

Fire Triangle: Fuel, heat and oxygen.

Flame: A rapid gas phase combustion process characterized by a self-propagation.

Flame Ionization Detector (FID): A nearly universal gas chromatographic detector., It responds to almost all organic compounds. An FID does not respond to nitrogen, hydrogen, helium, oxygen, carbon monoxide or water. This detector ionizes compounds as they reach the end of the chromatographic column by burning them in an air/hydrogen flame. As the compounds pass through the flame, the conductivity of the flame changes, generating a signal. This is the most commonly used detector in arson debris analysis.

Flame Propagation: Travel of a flame through a combustible gas/air or vapor/air mixture.

Flammability Limit: (See Explosive Limit)

Flammability Range: (See Explosive Range)

Flammable Liquid: A combustible liquid that has a flash point below 80°F according to the Coast Guard, 100°F according to NFPA. Liquids having a vapor pressure over 40 pounds per square inch at 100°F are classified as flammable gases. Flammable liquids are a special group of combustible liquids.

Flammable Vapor: A vapor/air mixture of any concentration within the flammability range of that vapor.

Flash Fire: A fire that spreads with unusual speed, as one that races over flammable liquids or through combustible gases. The temperature at which a pool of liquid will generate sufficient vapors to form an ignitable vapor/air mixture. The temperature at which a liquid will produce its lower explosive limit in air. Flash point describes one of several very specific laboratory tests. Frequently, materials can be made to burn below their flash point if increased surface area or mechanical activity raises the concentration of vapor in air above the lower explosive limit.

Fractionation: The separation of one group of compounds in a mixture from another, generally by distillation.

Fuel Oil: A heavy petroleum distillate ranging from #1 (kerosene or range oil), #2 (diesel fuel), up through #6 (heavy bunker fuels). To be identified as fuel oil, a sample must exhibit a homologous series of normal alkanes ranging from C_9 upward.

-G-

Gas Chromatography (also known as Gas Liquid Chromatography):. The separation of organic liquids or gases into discrete components or compounds seen as peaks on a chromatogram. Separation is done in a column which is enclosed in an oven held at a specific temperature, or programmed to change temperature at a reproducible rate. The column separates the compounds according to. their affinity for the material inside the column (stationary phase). Columns can be either packed or capillary. Packed columns employ a powder) substance which may be coated with a nonvolatile liquid phase. A capillary column is a glass or quartz tube coated with a nonvolatile liquid. Gas Chromatography (GC) is the accepted method for identification of hydrocarbon mixtures normally used as accelerants, and must be performed in order to have a valid identification of petroleum distillates.

Gasoline: A mixture of more than 200 volatile hydrocarbons in the range of C_4 to C_{12}, suitable for use in a spark ignited internal combustion engine. Regular automotive gasoline has a flash point of -40°F.

-H-

Headspace Concentration: A technique for concentrating all or most of the flammable or combustible liquid vapors in a sample on to a tube of charcoal, a wire coated with charcoal, a charcoal coated polymer, or some other adsorbing material which will later be desorbed in order to analyze the concentrated vapors. This is a primary form of sample preparation in arson debris analysis. This is also known as adsorption/elution, vapor concentration, or total headspace.

Heat: A mode of energy associated with and proportional to molecular motion that may be transferred from one body to another by conduction, convection or radiation.

Heptane: An alkane having the formula C_7H_{16}, flash point of 25°F and explosive limits of 1.2% to 6.7%.

Hexane: An alkane having the formula C_6H_{14}. Flash point of -9°F. Explosive limits of 1.2% to 7.5%.

Homologous Series: A series of similar organic compounds, differing only in that the next higher member of the series has an additional CH_2 group (one carbon atom and two hydrogen atoms) in its molecular structure. Fuel oils are characterized by the presence of an identifiable homologous series of normal alkanes.

Hydrocarbon: An organic compound containing only carbon and hydrogen.

Hydrogen: The simplest element. Atomic Number 1. Hydrogen gas has a specific gravity of 0.0694 (air = 1), so it is much lighter than air. Hydrogen is highly flammable, forming water upon combustion. Explosive limits are 4% to 75%.

-I-

Ignition: The means by which burning is started.

Ignition Temperature: The minimum temperature to which a fuel must be heated in order to initiate or cause self-sustained combustion independent of another heat source.

Immiscible: Describes substances of the same phase or state of matter (usually liquids) that cannot be uniformly mixed or blended.

Incendiaries: Substances or mixtures of substances consisting of a fuel and an oxidizer used to initiate a fire.

Incendiary Fire: Fire set by human hands.

Incidental Accelerants: Flammable or combustible liquids which are usual and incidental to an area where they are detected. Gasoline is incidental to an area where gasoline powered appliances are kept. Kerosene is incidental to an area where a kerosene heater is kept. Flammable liquids may also comprise a part of a product such as insecticide, furniture polish, or paint. Additionally, certain asphalt-containing building materials may yield detectable quantities of fuel oil components.

Infrared Spectrophotometry (IR): An analytical technique which utilizes an instrument which passes infrared radiation through a sample or which bounces infrared radiation off the surface of a sample. A very sensitive heat detecting device measures the amount of infrared radiation absorbed as the wavelength of the radiation reaching the detector is changed. IR can give useful information about the type of compounds present in a sample, but it is not capable of precisely identifying a complex mixture. Infrared is very useful in identifying single solvent acelerants.

Intumescent Char: In plastics, the swelling and charring which results in a higher ignition point. Used in the preparation of flame retardant materials.

Ion: An atom, molecule or radical that has lost or gained one or more electrons, thus acquiring the electric charge. Positively charged ions are cations; negatively charged ions are anions.

Isomer: One or two or more forms of a chemical compound which have the same number and type of each atom, but a different arrangement of atoms.

Isoparaffins: A mixture of branched alkanes usually available as a narrow "cut" of a distillation. Exxon manufactures a group of products known as "isopars" ranging from Isopar A through Isopar J. These solvent mixtures have a variety of uses. Gulf Oil manufactures a similar series of solvents, the most commonly available of which is Gulf Lite Charcoal Starter Fluid which is roughly equivalent to Exxon's Isopar G.

Isothermal: A type of gas chromatographic analysis where in the column is maintained at a uniform temperature throughout the analysis. (See Programming).

-K-

Kerosene (#1 Fuel Oil): Flash point generally between 100 and 150 degrees F. Explosive limits of 0.7% to 5.0%. Kerosene consists mostly of C_9 through C_{17} hydrocarbons. In order to be identified as kerosene, a sample extract must exhibit a homologous series five consecutive normal alkanes between C_9 and C_{17}. Kerosene is the most common "identical: accelerant, as it is used in numerous household

products ranging from charcoal lighter fluid to lamp oil to paint thinner to insecticide carriers. It is also used as jet fuel. K-1 kerosene has a low sulfur content required for use in portable space heaters.

Ketone: A type of organic compound having a carbonyl functional group (C=O) attached to two alki groups. Acetone is the simplest example of a ketone.

-M-

Magnesium: A silvery metal used in some metal incendiaries. The dust is highly explosive. Ignition point of 650°F.

Mass Spectrometry: A method of chemical analysis which vaporizes, then ionizes the substance to be analyzed, and then accelerates the ions through a magnetic field to separate the ions by molecular weight. Mass spectrometry can result in the exact identification of an unknown compound, and is a very powerful analytical technique, especially when combined with chromatography.

Meta-ethyltoluene (m-ethyltoluene): A component of gasoline.

Matrix: Substrate. The material from which a substance of interest is removed for analysis.

Methane: The simplest hydrocarbon and the first member of the paraffin (alkane) series; having a formula CH_4. Methane is the major constituent of natural gas. Methane has a heating value of 1009 BTU/cubic foot. Its explosive limits are 5% to 15%.

Methanol: Methyl alcohol. Wood alcohol. The simplest alcohol. Methanol is water soluble and has a flash point of 54°F and explosive limits of 6% to 36.5%.

Methyl Silicone: A nonvolatile oily liquid used in gas chromatography to separate nonpolar compounds. Methyl silicone columns typically separate compoundsaccording to their boiling point.

Methylstyrene: A common polymer pyrolysis product.

Mineral Spirits:' A medium petroleum distillate ranging from C_8 to C_{12}. The flash point of mineral spirits, sometimes known as mineral turps, is commonly used as a solvent in insecticides and certain other household products. Many charcoal lighter fluids are composed almost entirely, of mineral spirits.

Molecular Weight: The sum of the atomic weights of all of the atoms within a molecule. Generally, molecules of the same type have higher boiling points if the molecular weight is higher.

Molecule: The smallest particle into which a substance can be divided without changing its chemical properties. A molecule of an element consists of one atom, or two or more atoms that are alike. A molecule of a compound consists of two or more different atoms.

Monomer: The simplest unit of a polymer. Ethylene is the smallest unit of polyethylene. Styrene is the smallest unit of polystyrene.

-N-

Naphtha: An ambiguous term which may mean high flash naphtha (mineral spirits), or low flash naphtha (petroleum ether, low boiling ligroin) or something altogether different. Flash point and explosive limits vary. The term naphtha is so ambiguous that it should not be used.

Natural Gas: A mixture of low-molecular weight hydrocarbons obtained in petroleum bearing regions throughout the world. Natural gas consists of approximately 85% methane, 10% ethane and the balance propanes, butanes and nitrogen. Since it is nearly odorless, an odorizing agent is added to most natural gas prior to final sale.

Nebulize: To form a mist of fine droplets from a liquid. To atomize.

Nitrogen: A gaseous element which makes up approximately 80% of earth's atmosphere. Nitrogen is relatively inert and does not support either combustion or life. Nitrogen is usually found in the molecular N_2 form.

-O-

Octane: 1) An alkane having the formula C_8H_{18}. Flash point of 56°F. Explosive limits of 1% to 3.2%. 2) A measure of the resistance of a sample of gasoline to premature ignition (knocking). 100 octane fuel has the knocking resistance of 100% iso-octane (w,2,4-trimethyl pentane). Zero octane fuel has the knocking resistance of n-heptane. 89 octane fuel has the knocking resistance of a mixture of 89% iso-octane and 11% n-heptane.

Olefin: An alkene. An organic compound similar to an alkane, but containing at least one double bond. Olefins having the formula CnH_2n. The simplest olefin is ethylene, C_2H_4.

Organic Chemistry: The study of the carbon atom and the compounds it forms, mainly with the 20 lightest elements, especially hydrogen, oxygen and nitrogen. Some 3 million organic compounds have been identified and named.

Oxidation: Originally, oxidation meant a chemical reaction in which oxygen combines with another substance. The usage of the word has been broadened to include any reaction in which electrons are transferred. The substance which gains electrons is the oxidizing agent.

Oxygen: A gaseous element which makes up approximately 20% of the earth's atmosphere. It is usually found in the molecular O_2 form. Oxygen is the most abundant element on earth.

-P-

Pentane: An alkane having the formula C_5H_{12}, flash point of -40°F, and explosive limits of 1.4% to 8%. Pentane is frequently used to extract flammable or combustible liquid residues from debris samples.

Petroleum Distillates: By-products of the refining of crude oil. Low boiling or light petroleum distillates (LPD) are highly volatile mixtures of hydrocarbons. These mixtures are sometimes called ligroin, petroleum, ether, or naphtha. LPDs are used as cigarette lighter fluid, as copier fluid, and as solvents. Medium boiling petroleum distillates (MPD) are sometimes known as mineral spirits, and are used as charcoal starters, as paint thinners, as solvents for insecticides and other products, and as lamp oils. High Boiling or Heavy petroleum distillates (HPD) are combustible liquids such as kerosene and diesel fuel.

pH: A number used to represent the acidity or alkalinity of an aqueous solution. pH7 is neutral. Acids have a pH below 7, the lower the pH, the more acidic the solution. Bases have a pH above 7. The higher the pH, the more basic or alkaline the solution.

Photoionization Detector (PID): A type of detector used in chromatography which employs ultraviolet radiation rather than a flame to ionize compounds as they pass through. a detector. Photoionization detectors are particularly sensitive to aromatic compounds.

Polarity: The measure of an electrical charge on a molecule. Most flammable or combustible liquids are nonpolar. Many water soluble compounds,. including alcohols and acetone, are polar.

Polymer: A large molecule consisting of repeating units of a monomer. Polymers may be natural, such as cellulose, or synthetic, such as most plastics.

Programming: A method of gas chromatographic analysis which reproducibly raises the temperature of the column so as to allow better resolution of the components over a wide range of boiling points.

Propane: An alkane having the formula C3H8. Propane is the major constituent of LP gas. Explosive limits of 2.4% to 9%. One cubic foot of propane has a heating value of 2500 BTU's.

Pseudocumene: (1,2,4-trimethyl benzene) A component of gasoline.

Pyrolysis: The transformation of a substance into one or more other substances by heat alone without oxidation.

Pyrophoric Distillation: The slow drying and passive pyrolysis of wood materials.

-R-

Radiation: 1) Transfer of heat through electromagnetic waves from hot to cold. 2) Electromagnetic waves of energy having frequency and wavelength. The shorter wavelengths (higher frequencies) are more energetic. The electromagnetic spectrum is comprised of a) cosmic rays, b) gamma rays, c) x-rays, d) ultraviolet rays, e) visible light rays, f) infrared, g) microwaves and h) radio waves.

Resolution: 1) In chromatography, a measure of the separation of components. 2) In spectroscopy, a measure of the ability of the instrument to detect individual absorbance peaks.

Retention Time: The length of time required for a compound or component of a mixture to pass through a chromatographic column.

-S-

Saturation: The state in which all available bonds of an atom are attached to other atoms. Alkanes are saturated. Olefins are unsaturated.

Spalling: Destruction of a surface by frost, heat, corrosion, or mechanical causes. Concrete exposed to intense heat may spall explosively. Expansion and contraction of the concrete as well as vaporizing moisture contained in the concrete contribute to this effect. It does not necessarily mean an accelerant was used.

Spectrophotometer: An analytical technique devoted to the identification of the elements and the elucidation of atomic and molecular structure by measurement of the radiant energy absorbed or emitted by a substance in any of the wavelengths of the electromagnetic spectrum in response to excitation by an external energy source.

Spontaneous Heating: Also known as Spontaneous Combustion. Initially, a slow, exothermic reaction at ambient temperatures. Liberated heat, if undissipated (insulated), accumulates at an increasing rate and may lead to spontaneous ignition of any combustibles present. Spontaneous ignition occurs sometimes in haystacks, coal piles, warm moist cotton waste, and in stacks of-rags coated with drying oils such as cottonseed or linseed oil.

Styrene: Vinylbenzene. An aromatic compound having the formula $C_6H_5C_2H_3$. The monomer of polystyrene plastic. A common product of polymer pyrolysis.

Substrate: Matrix. The material from which a substance to be analyzed is removed.

Sulfur: A nonmetallic yellow element. A constituent of black powder, sulfur burns readily when in powdered form.

Terpenes: Volatile hydrocarbons which are normal constituents of wood.

Thermal Conductivity Detector: A type of gas chromatographic detector which is sensitive to the change in the ability of the gases emerging from the column to conduct heat. A thermal conductivity (TC) detector is not as sensitive as a flame ionization detector,' but is capable of detecting some molecules; such as water, which give no signal in FID.

Thin Layer Chromatography (TLC): A procedure for separating compounds by spotting them on a glass plate coated with a thin (about 0.01 inch) layer of silica or alumina, and "developing" the plate by allowing a solvent to move upward by capillary action. TLC is especially useful for identifying and comparing materials which are highly colored or which fluoresce under ultraviolet light. TLC is used extensively in explosive analysis and in the comparison of gasoline dyes.

Toluene: Methylbenzene. An aromatic compound having the formula $C_6H_5CH_3$. A major component of gasoline. Toluene has a flash point of 40°F and explosive limits of 1.2% to 7%.

Turpentine: 1) Gum. The pitch obtained from living pine trees. A sticky viscous liquid. 2) Oil. A volatile liquid obtained by steam distillation of gum turpentine, consisting mainly of pinene and diterpene. Turpentine is frequently identified in debris samples containing burned wood.

Vaporization: The physical change of going from a solid or a liquid into a gaseous state.

Volatile: Prone to rapid evaporation. Both combustible and noncombustible materials may be volatile.

-X-

X-ray Diffraction: An analytical technique used to identify crystalline solids by measuring the characteristic spaces between layers of atoms or molecules in a crystal. X-ray diffraction can be very useful in the identification of explosive residues.

X-ray Fluorescence: A spectrophotometric analytical technique which uses x-rays to excite the samples. X-ray fluorescence will allow the characterization of the elemental content of a sample.

Xylene: Dimethylbenzene. An aromatic compound having the formula $C_6H_4(CH_3)_2$. Xylene is a major component of gasoline. A mixture of toluene and xylene is frequently used as automotive paint thinner. Xylene is actually a mixture of three isomers, ortho, meta and para xylene, which have the methyl groups in different positions relative to each other on the benzene ring. The flash points of these isomers range from 81° to 115°F. Paraxylene, with a flash point of 81°F, is used to calibrate flash point testers. The explosive limits of xylene are 1.0% to 7.0%.

www.ingramcontent.com/pod-product-compliance
Lightning Source LLC
Chambersburg PA
CBHW081232170526
45165CB00009B/3047